FROM EAGLE TO EAGLE

A Boy Scout's Incredible Journey to Coast Guard Captain

From Eagle to Eagle: A Boy Scout's Incredible Journey to Coast Guard Captain
Smithfield, VA

Copyright © 2023 by Mark Ogle

ISBN 979-8-218-25690-6 (paperback)
ISBN 978-1-736-66010-2 (ePub)

Printed in the USA

27   26   25   24   23   1 2 3 4 5 6 7

The views expressed are those of the author and do not reflect the official policy or position of the Coast Guard, Department of Homeland Security (DHS) or the US Government. DHS cannot attest to the substantive or technical accuracy of the information.

The appearance of U.S. Department of Defense (DoD) visual information does not imply or constitute DoD endorsement.

All maps courtesy of Unsplash unless otherwise noted.

This is book is dedicated to those future heroes we call when it's all on the line.

To those who give their time and energy inspiring our youth.

To the great men and women of the world's Coast Guards and Navies.

To my shipmates and classmates, great friends, mentors and students from around the world who made my career one to remember

To my family. I'm so blessed.

Mark Ogle is known by many names and many accomplishments which are insightfully detailed in his Coast Guard journey. I proudly take credit for several nicknames, none of which adequately capture the immense contribution Mark has made to this nation. An accomplished expert in all Coast Guard missions and extraordinary leader, "Opie" has made Mayberry and this very grateful former Commandant very proud. Read this book and enjoy.

Thad Allen
Admiral, USCG (Ret)

Our Coast Guard Academy graduating class overwhelmingly selected Mark for the Superintendent's Award for leading the class in personal character, always pushing the boundaries to try new approaches and doing so with a unique brand of contagious enthusiasm that inspired his crews. You'll see that character shine through in his recounting of his many adventures in the service to our nation. Our paths intersected at many times over forty years, and Mark's exemplar leadership was always a tough act to follow but a privilege as well. Military leaders of all backgrounds—read, learn and enjoy.

Mike McAllister
Vice Admiral, USCG (Ret)

Mark Ogle represents a stellar example of the Coast Guard's Honor, Respect, and Devotion to Duty. I know firsthand how he demonstrates with passion and commitment in everything he does. This book provides an example of someone who loves what he does.

Vince Patton, Ed.D.
8th Master Chief Petty Officer of the USCG

I first met Mark at the Academy decades ago. He never stopped learning and honing his craft, and I recall fondly the many late nights we worked together, evaluating high-end counterterrorism teams, planning DOG operations, and fighting hurricanes and errant satellites. His exploits are legend. You're in for a great read!

Merrie Austin
Rear Admiral, USCG (Ret)

Table of Contents

Introduction

WHEN SPEAKING TO ANY audience, it's recommended that one gain their attention while simultaneously establishing credibility on the subject.

As an instructor for younger classes, instead of listing my postings—some of which were well before the students were born—I told them I would quickly list six career highlights, one of which was false. At the end of the list, I asked the class to pick the false one.

Of course, if you're a military person, you know how to make stories sound even more exciting than perhaps they were.

Here's the list I used most often:

1. I was the senior man for a Whitehouse-ordered, armed landing party on a Caribbean Island amidst gunfire to rescue over seventy barricaded tourists in advance of the 82nd Airborne Division.

2. A former commandant nicknamed me a *water retriever* because I swam down ten feet to pull a guy out of a submerged car.

3. I traveled to seventy-six different countries training foreign militaries, customs, and police, and I sailed across the Atlantic with the Navy's Second Fleet, bouncing from ship to ship to train boarding teams for Operation Desert Storm.

4. I commanded a unit that had an element on the ground in Iraq, played a critical role in the capture of America and Mexico's most wanted, interdicted the first drug-laden submersible, and seized a record-breaking 110 tons of cocaine.

5. I did an in-water protester mission in Hawaii that is both the subject of a book as well as a very funny *South Park* episode.

6. Originally a top-secret mission, I was chief of operations (J3) for the US government's fifteen-agency task force to successfully shoot down a spacecraft that threatened the planet.

Most will guess number six first, but that's actually true and covered later in this book. Usually, they guess number four last, which is the only one that is not *completely* true. Everything in it is correct except that we seized 130 tons of cocaine. But who's counting?

Bottomline, I'm an average guy from a good and humble family. I'm not a great student or athlete or even stunningly good looking. But I have managed to be in the right place at the right time and, most importantly, with the right people.

I had a unique front-row seat in some historical events. I will chronicle both the highs and lows, victories and mistakes, and highlight the role of leaders and shipmates along the way. I believe there are many lessons to be learned, especially for a young person embarking on a career. At a minimum, you will get a little geography, military rank, and international relations insights along the way.

I also wanted to tell my story for my family. My father died from cancer when he was my current age. He, like his father and many men I know, had trouble communicating with their family members. Because of what I did for a living, it was difficult and perhaps unwise to share details with my family. This is an opportunity for them to know what I was actually doing. It will span from Boy Scouts to commanding a patrol boat to commanding portions of three states during 9/11 to overseeing 3,000 special forces personnel conducting global missions.

This did not happen overnight.

Chapter 1

The Career That Almost Did Not Happen

THERE ARE MANY SIMILARITIES between Eagle Scout and a US Coast Guard captain. The insignia share our national symbol. The mottos are similar: *be prepared* and *always ready*. They also signify the end of a journey for most, consisting of six distinct promotion points, starting with the rank of scout and ensign. While I chose the officer route, many Eagle Scouts rose through the ranks in the enlisted corps and are some of the finest leaders I have ever met. Their insignia also includes the eagle.

As an eleven-year-old boy, I wore a uniform, was organized into a group, and followed clear specific requirements for promotion to the next rank for the first time. We had to memorize the scout law and take an oath. We had community service projects and, yes, adventures. We learned how to stand in formation, salute, and conduct ceremonies. With promotion came more leadership responsibilities. Scouts provided us with role models outside of the family including Scoutmasters, camp leaders, and boys a few years older than us.

None of this would have been possible without adult volunteers. Those individuals camped in the rain and snow, hiked mountains, paddled rivers, and taught wilderness skills instead of staying in the comfort of their homes. For those currently giving their time and energy to raise our youth, I salute you. You are fulfilling our generational responsibility.

Every generation is different and growing up as a kid in the '70s looks very foreign to today's youth, just ask my son or daughter. But with age comes wisdom, and we must question if our rapid technological advances truly improve life, liberty, and the pursuit of happiness. When you think about it, while we've achieved a very comfortable lifestyle, life's greatest moments are generally when we are least comfortable: getting married, having a child, sky diving, or playing in a championship game. Sometimes you take the leap knowing you may fail, and that's okay–*except in skydiving*.

Before diving into military missions, here's a little context about me. I come from a family of three boys; all have managed to be successful, yet all took different paths. One succeeded in the public sector, one in private, and one in both public and private. One thing we did have in common, we all earned the rank of Eagle Scout before leaving the nest, which we did before the age of 18.

My older brother by a year is Dr. John W. Ogle III. We just called him Johnny growing up. He was a high school valedictorian (number one out of 400) and went to the Air Force Academy. He was number one out of 1,600 cadets in his "doolie year" but slacked off a bit (his words) and finished 23 out of 1000. After serving his payback tour of five years, he left active duty but remained in the reserve while going back to school to become an emergency room physician. He went to medical school at University of North Carolina (UNC) and did his internship at Stanford. In 2000, he was named Flight Surgeon of the Year for doing the first-ever, full cardiac resuscitation for a patient on the continent of Antarctica. Gary Dunckel, who was the patient, survived and later attended Johnny's wedding. As I write this, Johnny is deployed to the Middle East, literally taking care of our troops during the COVID crisis. Johnny has never had children but is a great uncle. Both younger brother Scott and I eagerly ship our kids to Colorado for heart thumping adventures.

I'm the middle child but was the youngest for ten years. My little brother Scott, who we often referred to growing up as the "mistake," certainly did not turn out that way. He, like Johnny and me, was a particularly promising student and was selected for an elite high school program focused on science and math. He effectively left the house when he was sixteen. Upon graduating high school, he chose UNC at Chapel Hill, married his childhood sweetheart Jenni, and has three beautiful girls.

Bravery is often associated with first responders and the military, but I will say he is one of the bravest people I know. He made big gambles right out of college. He moved to New Jersey and then out to Silicon Valley before returning to North Carolina, where he hooked up with a small start-up software company called Sage Works. Then with his young family, he moved to London to launch the company's products overseas. The company's founder was impressed and made Scott the CEO. The company flourished under his leadership, growing from a handful to 250 people. Other companies took notice.

Scott is a risk-taker and an entrepreneur. He worked extremely hard through nights, weekends, and holidays. Jenni said she rarely saw him in the first sixteen years of their marriage. In the end, the company was sold, and Scott had a multi-million-dollar payday. He has been able to retire in his early forties with a lavish family lifestyle. Now he does all sorts of charitable things with his time and resources, including raising $100K for the local Boy Scouts, which had such a huge impact on all of us.

That just leaves me—the middle child with overachieving siblings. Unlike my brothers, I faced some medical challenges at birth, with the chance of survival at twenty percent. I beat those odds but, as you can imagine, following a year behind Johnny was no easy task. I suspect our competition made us both better.

Ogle family photo, circa 1992. From left to right: Scott, Mark, John (Dad), Ruth (Mom), and Johnny

Our dad was a math teacher, so he knew education and, unfortunately, knew our teachers. His father, John W. Ogle Sr, retired in New Orleans as a Coast Guard chief petty officer whose career included service during WWII. Growing up, we made the bus ride from Raleigh to New Orleans, where our granddad told us the Coast Guard's stories. He had tried to get into the Coast Guard Academy but didn't make the cut. My godfather, who was my dad's brother Louis, also went to Coast Guard Boot camp but, unfortunately, did not make it through due to a medical issue.

Our mom, Ruth, was a stay-at-home mother and a very accomplished artist, earning state recognition. She produced many successful art shows and workshops. Because she was the primary nurturer and helped me specifically through early medical challenges, we have a special connection. Mom came from a very humble family growing up in the depression in New Orleans. Her family raised chickens in the backyard just to get through lean times. As a kid during World War II, she picked up scrap metal for the war effort. Their small house in Metairie did not have indoor plumbing either. She and her four brothers took turns dumping bedpans in the outhouse.

Fast forward a decade to 1954 when one of her brothers got a job working for the city. He used his connection to get Mom into a high society competition. Mom, at the age of nineteen, was crowned queen of Marti Gras. This was a huge honor, and she traveled and even met the US Speaker of the House. But she never lost her humble roots. She taught us to be thrifty, not to complain, always be kind to people, and capitalize on opportunities.

I entered Boy Scouts with two of my best friends, Chip Shankle and Frank Cope. For me it was like an extended family. I enjoyed the camaraderie of weekly meetings and monthly camping trips, regardless of the weather. One year, I decided not to use a tent for the entire year, which meant eleven nights in leaky lean-tos and one in a snowball igloo that turned out to be the coldest night of my life!

As we got older, we faced more difficult challenges. There were overnight canoe trips and the Order of the Arrow ordeal, which amounted to a program with Spartan-like events. The weekend camping trips became summer week-long, fifty-mile, backpack trips in the mountains of North Carolina and Virginia. I remember having to lie about my weight so I could go. You see, for the fifty-miler, you had to weigh one hundred pounds. I was in the low nineties, and my pack weighed forty-five pounds. It was brutal, but there was nothing like finishing, despite rubbing all the skin off my heels.

We completed a one-day, fifty-mile hike at Fort Bragg. We started around three p.m. as it began to cool off and walked for seventeen straight hours with only a few water breaks. Fort Bragg was abuzz with war games throughout the warm summer night, using tracer rounds and aircraft flying overhead. Our small group was one of the few that finished, and we even ran the last mile. For the record, the last ten miles felt like the first forty.

I guess I liked the structure of scouting as Troop 357 had fifty boys organized into patrols and led by a senior staff of older boys. Our head Scoutmasters Joel Reams, Jim Partington, Mike Matzinger, and Don Davenport were truly selfless in giving their time and attention. While he didn't talk about it, we knew Don had been in Korea and, at one point, was in hand-to-hand combat. Sporting a flat gray crew cut, he was a stern yet compassionate guy with little sympathy for those complaining about our often-uncomfortable environment.

On one camping trip, a small group of us moved into a heavily wooded area, searching for a good campsite. Dripping with sweat, we were eager to get our packs off when we heard a younger kid behind us screaming. When we looked back, the kid was engulfed in a cloud of angry yellow jackets. These insects burrow in the ground, and he must have stepped on the nest. By killing a couple, it released the pheromones that made them crazy.

The yellow jackets came after the rest of us, so we ditched our packs and started running like we were being chased by a cheetah, swerving and swatting to a distant lake. By the time we reached it, most the wasps were gone, but the boy, probably twelve years old, was in bad shape. He was covered in welts and broken off stingers. I suspected he had been stung maybe one hundred times. He was crying, and we helped pull the stingers from his hair, socks, and even his underwear. The few that were still buzzing around kept going after this poor kid. We killed as many as we could. Then he told us he was allergic to beestings, but his medicine was in the pack now perhaps a half mile away. He was starting to feel quite ill.

Based on what we learned from the first aid merit badge, we knew he might be going into shock, and his airway could be in jeopardy. Off in the distance at the edge of the lake, we saw a building and a car. Another scout and I carried the kid at double time to an adult with a car, and they sped off to the hospital.

We were able to cautiously return to the sight and retrieve all the packs. We didn't camp there!

The next day, the kid was back, and his parents wanted to thank us personally. We found out that, due to his current condition and the sheer number of stings he received, it was a good thing that his medicine was unreachable and not administered. Risking more stings during a retrieval and delays getting to the emergency room would have led to a far worse outcome, according to the doctors at the emergency room.

I have to say this was probably my first real rescue, and I liked it! I also realized that I needed to start lifting weights.

It was all good preparation for my greatest childhood adventure with Chip Shankle. We were probably fourteen years old when we boarded a bus full of scouts with a couple of adult leaders for the long drive to Cimarron, New Mexico. The high adventure camp called Philmont awaited us. There, we backpacked seventy miles, the highest point being the snowcapped peak of Baldy Mountain that rose over 12,000 feet.

Chip Shankle celebrates completing seventy-mile backpack trip in New Mexico at Philmont

In summer camps in North Carolina, we earned merit badges like wilderness survival, environmental protection, lifeguarding, citizenship, and even motor boating. Different badges were required to promote to certain ranks. On the path to Eagle Scout was the position of 1st class. Back then, for that rank, you had to earn a swimming merit badge. It was, I'm embarrassed to say, an issue for me. You see, I had a fear of putting my head underwater, not to mention swimming. After the swimming badge, we still had to earn a lifesaving merit badge before making Eagle

Scout. These badges were not learned in a pool either; it was a murky lake. We had to swim down ten feet for lifesaving, then locate and bring up a cinder block. With a body mass index in the single digits, my inherent buoyancy was similar to the cinder block target.

I knew I had to face my fear. I started slowly in the bathtub then moved to a local shallow creek. I could lay on the bottom of the creek for over a minute without coming up for air. Skinny kids don't float. I eventually built confidence and earned those merit badges and advanced. I used far more of these skills as an adult than what was taught in school. That included recovering a man trapped in a submerged car.

I knocked out the requirements, advanced, and became a patrol leader. Our patrol—self named the Golden Eagles—consisted of eight boys. In competition with five other patrols, we won the head-to-head competition for the year based on advancement, fitness challenges, and attendance. That was a great leadership laboratory for me. The troop eventually elected me as the senior patrol leader, which was the scout equivalent of

Sample of scout badges to earn

the top dog. This experience taught me initially to pull my own weight, work well in small groups, then motivate others, and accomplish tasks. It also taught me that if you're given the authority and responsibility of a leadership title, you will also be held accountable for the group's performance, good or bad.

Boy Scouts and decent grades alone were not going to cut it for the future I envisioned. I wanted no part of living an average boring life. I was probably fourteen or fifteen when I started the "grand initiative" to pad my resume. I managed to get elected to the student council in high school. I also knew I must have an athletic letter. I was too light for football; besides I did not have the required parental blessing. My best friend Chip was huge—weighing in close to 300 pounds as a senior—and was surprisingly fast. He was honored as our class's top athlete for accolades in both football and wrestling. I was honored to have Chip as my best man when I got married.

Just knowing a great athlete won't earn you a letter. Based on my father's recommendation, I pursued track. Although I must say, I have never really enjoyed running—ever—nor was I good at it. I eventually lettered in distance and hurdles. If nothing else, the coach said it would get me in better shape, and it did.

I was also eager to gain independence early, so in the summer of 1980, I applied to the Youth Conservation Corps for an eight-week program in the mountains near Boone, North Carolina. This co-ed program of about thirty teenagers was based on the campus of Appalachian State University. We were berthed in the dorm and did various manual labor jobs for the National Park Service. The work included cutting trails, building campgrounds, backpacking trout into the mountains to stock streams, and putting up cattle fences. Along with being 100 percent self-sufficient, I officially had a girlfriend and pocketed $500. That money served as a down payment for a used, red Camaro.

My brother Johnny had always been interested in flying and had joined the Civil Air Patrol led by retired Air Force Lieutenant Colonel Hawkins, who also happened to be a high school principal. I followed a year later and served on the Civil Air Patrol's Ground Rescue team. I even deployed to a small plane crash in the woods near Raleigh Durham's airport. At night in the rain, I saw a horrific sight as the plane and its occupants hit the woods at well over one hundred miles per hour. While no one would want to see what I saw that night, it better prepared me for future military missions, including responding to downed aircraft in the water with a young crew.

After high school graduation and just before reporting to the Academy, I went on a 180-mile canoe trip on the Neuse River with the Civil Air Patrol and made it all the way to the beach. Toward the end, we lashed canoes together and fabricated sails. One of my favorite books growing up was Tom Sawyer, and I think this marked a fitting end to my childhood. I also think all that extracurricular activity helped compensate for my less than stellar SAT scores.

As a high school junior, I watched my older brother Johnny apply and get accepted to the Air Force Academy. I remember him packing up and heading to Colorado—no more reliance on the family. My dad was not a fan of the military despite, or perhaps because of, having been a military dependent child. When I also expressed interest in the military, he tried to dissuade me. He went so far as to say I would not make a good military officer. Paradoxically, Dad's lack of encouragement kindled my rebellious spirit somehow. It motivated me to pursue my own destiny further. I took the initiative and applied to UNC, Appalachian State, and the Coast Guard Academy. My best friends, Chip Shankle and Frank Cope, both decided to go to Appalachian State. I spent the summer on that campus with the Youth Conservation Corps just two years earlier and knew it would be a great option.

Then there was the money issue. I did not want to be beholding to anyone. So, I applied and got a full-ride Army Reserve Officers' Training Corps (ROTC) scholarship to Appalachian State. My other close friend from the neighborhood was Peter Caliendo who wanted to fly like Johnny and headed to Embry Riddle University in Florida. Everyone was leaving the area! I did not have the passion or coordination to be a pilot despite passing the aviation test. My aircraft experience in the future would be as a passenger, sky diving, and fast-roping.

Then came the letter in February of 1982. It read, "Thanks for your application to the Coast Guard Academy. You are considered an alternate." Meaning, if their preferred choices decided not to attend, maybe, just maybe, I'd have a shot. But it would be late spring before I would hear anything. I probably should have been happy to make the back-up list, but I was bummed as this was my first choice by far.

Humbled, I focused my attention on UNC. It was in the town where I was born and where my father had gotten his master's degree. The only problem: Chapel Hill was only thirty minutes from the house. At the time, UNC was considered a particularly good school (and still is). I was familiar with the Tarheels as they were the archrival of the Wolfpack of Raleigh's N.C. State, where I sold drinks in the stadium for both football and basketball games. I took breaks to watch Michael Jordan play when the Tarheels came to town.

I was accepted to Carolina and landed a full-ride Navy ROTC scholarship. I even went so far as getting a dorm room.

A second letter from the Coast Guard Academy came in late spring. I vividly remember sitting in my parents' car in the garage and slowly opening that letter. I was in! I honked the horn for several minutes, probably freaking out my mom and the neighbors.

At the Athens Drive High School Class of 1982 pre-graduation award ceremony, students were recognized for scholarships, including their value. The student body practically passed out when they heard some of these gifted students received as much as $10,000 scholarships. Back then, that was real money. I was an average guy, friendly, but did not run with the most popular crowd. It was not cool to be in the Boy Scouts back then either. We had code like "are you going to the function this weekend?"

I was one of the very last to be recognized. When I was called forward to receive the acceptance letter to the Academy, the Coast Guard Auxiliarist who read it talked about sailing aboard the *Barque Eagle* and the low acceptance rate. When he stated the scholarship was valued at $160,000, it stunned the crowd. The silence gave way to very enthusiastic applause. I was representing them.

I missed the prep sessions and the programs designed to acclimate new Academy students, but my brother Johnny had given me a good idea of what to expect in my freshman year, so I started running even more and doing push-ups—lots of push-ups. The two-week paddle to the beach probably helped as well. I could easily do one hundred straight push-ups when I left. Good thing, as we did a thousand a day that first summer.

We also had square meals for the first year that had nothing to do with nutrition. It actually meant sitting on the front three inches of our chairs and lifting the food straight up until it was eye level, then bring it straight back to our mouth. It's a wonder I gained any weight.

It was not going to be easy.

Lessons Learned:

Do not let fear hold you back; learn to swim.

Do not let others tell you that you can't succeed.

Carry your weight and then some more for those needing help.

If it's hard, it's probably worth doing. Always finish, even if it is in the last place.

Treat others as you want to be treated. Put yourself in other's shoes.

A few great friends are better than many acquaintances.

Surviving in the wilderness with no food or shelter does two things: builds confidence and gives you an appreciation for what you have.

Be a force for good in the world, which means defending the innocent and confronting evil.

Chapter 2

The Academy

THE TEN-HOUR DRIVE NORTH to New England was eye-opening, especially for this southern boy. It was a lot cooler than North Carolina. It wasn't just the temperature that was cool; it was the reception. Apparently the good folks of Connecticut and our upperclassmen didn't appreciate a *Mayberry-style* welcome. I called those folks *mass-holes*.

The first morning, I was dropped off at 0800 in front of Chase Hall which is the barracks for all cadets. I had long, bleached blonde hair from my long canoe trip. That must have irritated the juniors who served as our cadre. These rope-wearing upperclassmen whisked all the newly reported freshmen into the courtyard and out of

Typical freshman learning moment at the Academy (photo courtesy of Coast Guard DVIDS)

sight of the parents. It was on. By early afternoon, we had no hair, were already in a uniform held together by safety pins and were marching in front of our parents.

The first summer was a physical, mental and emotional challenge. Like all boot camps, it's designed to tear you down and make everyone equal, then slowly build you up to be part of an effective team. Our memories were constantly challenged. We had to study the newspaper to know the movies in the surrounding five cities although freshmen did not go to a single film that the first summer. We were also required to always have the next three meals committed to memory.

Upper-level students repeatedly quizzed us from reveille at 0600 to taps at 2200. During waking hours, our doors had to be open at ninety degrees, so there was ample time for "engagement." There were penalties for not knowing an answer. Beyond the old reliable push-ups, these guys were highly creative. *Flight ops* that meant running with a full seabag up and down flights of stairs. *Golden arches* required us to bend over backward to place our noses on the top bunk. *Bracing up* meant contorting so our neck's back was flat against the wall, which created multiple chins in the process. On the plus side, I think I got an inch taller.

Then there was holding the lead-filled rifle straight out in front of us for upwards of twenty minutes. But those creative upperclassmen were not done. There were *green benches* where we sat against the wall with our backs flush and legs at ninety degrees. These would get our thighs burning. For fun, they also invented *red benches* that was a standard green bench but with a bayonet below our butts. While we are undergoing these educational training sessions, we were also quizzed on other things like characteristics of ships and aircraft, mottos, and creeds. We were not going to get to complain to mom and dad either. After the march on the first day, contact with the family ceases.

I think my worse moment of that first summer was when I went to the post office with two classmates. We grabbed a haircut, hit the mailroom and the snack dispenser. When we headed up to the barracks an hour later, we were greeted with a very unpleasant sight. As we squared the comer in the center of the passageway, our entire platoon was in front leaning rest, which is the push-up position. A river of sweat coated the deck as it was nearly one hundred degrees that day in late July. Our platoonmates were red and shaking. We froze in our tracks. We were the unaccounted-for cadets who had brought on this wonderful forty five-minute, character-building session. We immediately dropped to join the platoon, but our upperclassmen didn't allow it.

Standing in the shimmering light of a pool of sweat, one of my companions had to describe his haircut slowly, the other had to read aloud a rather personal letter from his girlfriend, and I had to eat my Twinkie and describe it to everyone. Needless to say, this was not a top bonding moment for us with the rest of the platoon.

I will not give up all the freshman year's secrets, but I will say the military stuff was relatively easy for me. It's a game, I knew it, and I played it. They applied pressure and tried to break us to ensure we would not crack later when it really counts. No one was going to break me. The cadets who had the most trouble had never backpacked fifty miles as a Boy Scout.

Despite a rigorous screening process, only about fifty percent of the class makes it to graduation. Some determine the lifestyle is not for them, some leave on demerits, and others for not passing the semi-annual fitness test or perhaps an honor code violation. But the vast majority leave due to academics. I had grown up in North Carolina, the national epicenter of academic excellence. To put this in perspective, I was in the top three percent of my high school class. Here, I was doing everything I could as a freshman to stay above 2.0.

I finally hit my stride in the middle of my sophomore year. Except for the first summer, the summer programs were great. We went to flight training in Mobile, Alabama, where I learned to land a helicopter on the water. (It was an amphibious helicopter in case you were worried.) We sailed around New England on a forty-foot sailboat, but the pinnacle was sailing the 295-foot Cutter *Eagle*, which was a Nazi war prize from WWII.

Coast Guard Cutter Eagle
(photo courtesy of Coast Guard DVIDS)

For the *Eagle*, we flew to Puerto Rico in a Coast Guard C-130 and sailed the cutter home via Bermuda. Coming through the old Bahamas channel, we were hit by a squall in the middle of the night. There's nothing like being in a deep sleep, waking up to alarms, and standing at a thirty-degree angle with the ship heeled over. The crew piped over the intercom for us to report to sail stations, which meant all hands on deck, literally.

We had to get aloft in the rigging quickly and methodically to bring in the sails before the wind pivoted to the bow and forced the ship backward.

Once on deck, despite the darkness, I saw we were flying through the water, trailing luminescence in our wake. The *Eagle* was so heeled over that the yard arm on the main was just feet above the water. Quite scary as we were only locked in the safety harnesses when not climbing. It was 150 feet up to the top of the mast, but climbing

Cadets manning sail stations aboard USCGC Eagle
(photo courtesy of Coast Guard DVIDS)

was generally better than the tug of war sessions on deck. Of the five sails on the front two masts, the best—in my opinion—were the Royals, which were the uppermost.

Eventually we were able to methodically pull in enough sails to slow the *Eagle* to a manageable speed.

While I liked sailing, I preferred going on regular cutters during those cadet summers doing actual missions like search and rescue and law enforcement.

Back at the Academy, because we had fewer than 1,000 cadets and competed with much larger New England schools, every cadet had to be in at least two sports every year. I played intramural football and intercollegiate rugby, among other sports. On the rugby team, we traveled to Annapolis to play the undefeated Naval Academy. They had multiple teams, TV cameras and numerous coaches. They even had Napoleon McCallum, a football running back using rugby to stay in shape. He was a Hall of Famer and went on to play six seasons with the Oakland Raiders. The Naval Academy players were admittedly very intimidating. We were more like a club team that focused more on keg parties than rugby, but we went out and scored on them first. That must have pissed them off as they put a whooping on us after that.

Then there was the matter of Norwich Academy. If Annapolis had West Point as a rival, we had the Connecticut-based, private, army school. Every year they tried to sneak aboard the Academy in the middle if the night and steal our mascot—a healthy eighty-pound black bear cub named Objee. They paid a high price during my freshman year as apparently Objee didn't care for their car's upholstery. We caught them before getting off the campus. Restrained in handcuffs, in front of the corps of cadets, they enjoyed a Coast Guard breakfast delicacy of cream beef on toast affectionately known as "shit on a shingle."

Just as in high school, I made several close friends in the Academy. Together, we took advantage of the military's Space-A program that allowed us to go to Air Force bases to see where the planes were flying. If they had room, we flew for free. It was an excellent way to see the world on a student's budget. My good buddy J.J. Kauza and I flew to the Azores. Geoffrey Trivers and I made it to Florida for spring break, and Terrence Keenan, Greg Bush, Pat Geddes and I traveled to Europe.

Classmate Eric Vernon and I after solo skydiving

Coming out of the summer, rising seniors were assigned leadership positions in the Corps of Cadets. Similar to the Boy Scout organizational structure with patrols and troop leadership, we had squads, platoons, companies, battalions and the regiment. Despite a high military cadet rank, my grades only placed me in the top two-thirds of the class, so I was not confident about getting a good job. Then, one of my best friends and fellow summer platoon commander had a bump in the road.

He got into trouble for being too aggressive with his summer platoon. I think they were reenacting the Normandy invasion at the Thames River's waterfront, which didn't sit well with the commandant of cadets. My friend was slated to be the Bravo Company Commander but lost the opportunity over the incident. Mike is now a Coast Guard vice admiral and my longest serving classmate in uniform, so those events didn't hurt him too badly. The incident probably helped him when he was later acting commander in New York City during 9/11 and even briefed President Bush. More on Mike later.

The company officer was looking around for a candidate to fill the position. I offered that I wasn't doing anything particularly noteworthy and was available. So, by luck, I landed the job. But it almost ended as soon as it started.

You see, there was a pep rally before our very first football game. Each company had its freshmen design a banner with the juniors providing oversight. Admiral Nelson, the Academy's superintendent, and his wife. sat in the auditorium's front center row. Our proud artists then unfurled the Bravo Company banner. It said something to the effect that we will cut off your balls . . . with a drawing of a bleeding football.

Instead of the audience cheering, there was much whispering—and not the good kind. I sank in my seat. When the rally concluded, I was summoned to the admiral's office. Mind you, I had never met him or been in his office. If you saw the first *Top Gun* movie where Maverick and Goose buzzed the tower, this was my version of it. Bottom line: his message was crystal clear: if you are in command, command. Trust but verify. I was on notice.

I sprinted back to the barracks as my company lined up on the roadway for the competition drill that preceded the first football game. I scrambled to put on my uniform and sword belt. There were eight companies lined up on the road in alphabetical order, so I had to pass all but the Alpha Company. Everyone knew I had just gotten my butt reamed by an admiral. I made haste passing interested onlookers. There was a row of academic buildings between the line-up spot and the parade field, where all the drill graders were stationed. When I finally got to Bravo Company, it was not the normal reception. All were silent; eyes on me. It was a pivotal moment, and I knew it.

You see, Bravo Company in corps competitions for the past couple of years was always in a dog fight for seventh place out of eight companies. When I looked out over the one hundred cadets this time, something appeared different. Perhaps it was the infusion of the freshman class. *USA Today* had just rated the US Coast Guard Academy as the most selective college in the country, besting the other military academies and Ivy League schools. The class of 1989 were the best of the best—

except when it came to banner artwork. Gary Thomas nudged me to say something.

I'm not a fan of public speaking, but I spoke from the heart and said only four words, "I'm tired of losing," and then did an about-face.

There was silence for several long seconds, and then it happened. In the very back where the banner artist freshmen were mustered, they started tapping their rifle butts on the street. More and more joined in, and the tapping became pounding, and then they started chanting, "Bravo."

It got louder and louder as the other companies looked on, and I'm told it carried over the building and onto the parade field. I must say, even though it was just a parade, there was something about stepping onto a field with one hundred people behind me with fixed bayonets. It was like a *Braveheart* moment without the face paint. I was focused, and they were too. When we marched off, I called out, "Eyes right."

My eyes locked on Admiral Nelson. He gave me a nod and a smile. He knew he had set me up.

Back in Chase Hall, the intercom came to life. I knew they only announced the top three companies in the drill competition. Usually, everyone was going about their business, changing our uniforms for the football game, and not listening. Why should they, as I don't remember ever placing over the past three years. This time it was eerily quiet. In third place: Foxtrot Company. In second place: Hotel Company. Then there was a pause, almost as if it had been a mistake. In first place: Bravo Company!

The hallways exploded with cheers. We're not losing anymore. I had their back, and they had mine. From then on, I had to come up with new, short phrases like, "They said last week was a fluke."

We won again and again and took the whole semester-long competition that included sports and other events.

During that fall, one of my best childhood buddies who had gone to Embry Riddle met the love of his life. Pete Caliendo asked if I would be his best man. Of course I was honored and agreed. The wedding was set for the summer after we both graduated. When I got home at Christmas, just six months before the ceremony, my parents had some very tragic news. They had withheld it for a few days as they knew I was engrossed in final exams. Pete had been in a plane crash and had been killed. I had not dealt with death before then. It hit me extremely hard. I didn't know how to engage his family or the fiancé I had never met. That was something I knew I needed to work on. Pete was a great guy, struck down far too early in life. His death taught me to live in the moment as you never know when your number will come up.

To this point, I traveled mostly with guys. After the holidays and the terrible news about Pete, I was approached by a young lady from the junior class. She knew only seniors were allowed cars, and there were four pretty ladies eager to go on spring break in Florida. They even had a house rented. I had been saving and had purchased a used red Camaro. It was already paying off as I agree to the plan.

The only catch was I had to take the challenging deck watch officer exam and score at least 90 percent. Back then, every graduating cadet was required to go on a ship as their first officer tour. Passing the deck watch officer exam during our senior year helped us qualify quickly once reporting aboard.

As a motivational tool, the school policy stated that whoever did not pass the exam lost spring break privileges and must remain at the Academy for an extensive study program. Talk about incentive! I studied my butt off and scored an unprecedented 100 percent. Even though I was a chauffeur, it was nice being surrounded by ladies for a change.

The Academy forms lasting bonds; at our thirty-year class reunion, over 50 percent of our class showed up. Some traveled as far away as Singapore.

Since the statute of limitations has passed and both have retired, I can now spill the beans that 1st Class Cadet Gary Thomas was secretly dating the lovely Ensign Cari Batson, who was two years our senior. For the record, you aren't supposed to do that. However, it was true love. She resided off campus, and I knew Gary jumped the fence. If we had an accountability drill, I called him (never lying) and stalled as best I could while he ran back. Gary asked me to be his best man in the wedding to be held shortly after graduation. He and now-retired Rear Admiral Cari Thomas married in Florida. Ever the popular partier with questionable military bearing, Gary had the whole posse down for the wedding. Gary's mom still holds me responsible for the epic bachelor party. A learning point: don't play tackle football in white uniforms near a pool.

Our senior year was full of anxiousness. I was happy to move out and do actual missions, but my selection priority was based on class rank.

Class rank was calculated 25 percent military and 75 percent academic. So, I was down the list pretty far.

Over the winter, the Academy brought in guest speakers to give us a sense of what life would be like after graduation. Admittedly, I slept through some of these, but not LTJG Dan Albright, a patrol boat skipper who had made numerous drug busts in the Caribbean. This was right about the time Miami Vice was dominating TV. I wanted to be *that guy*, so I needed a boat down south.

Coast Guard Cutter Valiant
(photo courtesy of Coast Guard DVIDS)

Luckily, I was able to get to the Cutter *Valiant* out of Galveston, Texas. I was selected with two great classmates Hung Nguyen and Kevin Orloff, as well as LCDR James Loew, one of our Academy instructors. He was going to be the XO. After we received our assignments, I asked him in front of the class if I could call him Jim. It was a joke, of course, but I thought he might kill me. He became one of my absolute best mentors.

Just months before graduation, I learned a second particularly important lesson on following the chain of command. Technically, a cadet is considered active duty military. Many of the seniors watched the news and saw a Russian fishing fleet trapped in the ice. It became a desperate situation. At the time, the cold war was still very much alive. As a US citizen, I wrote a letter to the president, much as I had done as a Boy Scout working on citizenship merit badges. In the letter, I recommended that he publicly offer our Coast Guard Ice Breakers to support a rescue mission. I went on to say, should they refuse the offer, they will look bad.

If they accept, the United States will look like we have offered an olive branch. Plus, we would get high marks internationally for trying to do the right thing. I mailed it off knowing full well it would not make its way to President Reagan's desk.

About a month passed by, and I got a reply. It was a letter from the White House to the commandant to the superintendent, through the commandant of cadets, through my company officer. The response was, "Thank you for your concern. Please follow the chain of command next time."

Finally, it was the Friday before the Saturday morning graduation at the Coast Guard Academy. I had survived the four years. I had my family up plus my Coast Guard-veteran grandfather and grandmother.

It became one of my absolute best and very worst days of my life. That morning there was an awards ceremony. The Superintendent's Award was given to the cadet, who, in the opinion of his classmates, led the class in personal character. When they announced my name, I passed my cover to my little brother Scott, and went on stage to get a silver bowl. I received a standing ovation from the class of 1986. It certainly was a wonderful moment provided by my class for the whole family, especially my grandfather, who spent a career in the Coast Guard.

That night we had a big dinner hosted by the Academy. My grandfather left the table to use the bathroom. It had been a while, so my father and I tracked him down. We found him unresponsive in a stall. We scrambled to get an ambulance, and I drove my grandmother to the hospital, but it was too late. The next morning, my older brother Johnny flew in and filled in for my grandfather to present my commission. Looking back on it, I'm pleased that my grandfather's final moments were with his family and the service he loved.

Lessons Learned:

The best speeches are usually *real* short.

Regardless of what they say, everyone wants to be on a winning team.

Life's low points can turn into life's high points, and vice versus, in a truly short period. Stay positive and surround yourself with good people.

Setbacks happen. They are learning opportunities. What does not kill you makes you stronger. - Conan

When it really counts, study hard.

Good character is being trustworthy, loyal, helpful, friendly, courteous, kind, obedient, cheerful, thrifty, brave, clean, and reverent. (That's Boy Scout law, by the way.)

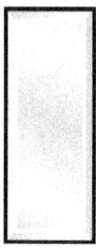

Chapter 3

My First Ship

THE BEST PART ABOUT making ensign or earning *butter bars* is no more calculus!

The Academy was in the rearview mirror, and I was officially an officer—on paper. Screwing up is expected, and ensign is considered a learning position. No doubt, I tested the command's patience. As long as an ensign has a good chief's mess looking out for them, it would have to be a pretty severe screw up not to make the next rank in eighteen months.

The pay was great compared to the meager allowance received as a cadet. As a freshman cadet, I learned to live off $40 a month. But everything had been provided, from meals to accommodations. There were no real bills.

Those who skipped college and went right into the workforce had a big advantage in building life skills and independence. Academy graduates were behind the times. The weeks following graduation were spent renting an apartment and accumulating life's necessities: inflatable furniture, pots, pans, and even a waterbed. We were not proud. When we got the intel there was a nice-looking sofa out by the dumpster, we'd grab it. Simple things like paying the bills when you're at sea or getting regular oil changes for your car had real world consequences if neglected. Academics were over, and real world learning more than filled the void.

After the emotional roller coaster of graduation and my grandfather's passing, I think the long drive to Galveston, Texas, was helpful. I had been to Texas once before on the Boy Scout bus ride to Philmont, New Mexico. My brief impression of Texans: they were confident, patriotic and friendly; exactly what I found in Galveston. I transitioned from teenage, dependent, Academy life to early twenties independence by sharing an apartment with Kevin and Hung, my classmates.

Coast Guard Cutter Valiant underway
(photo courtesy of Coast Guard DVIDS)

My first ship was the 210-foot, medium-endurance, Cutter *Valiant*. It was already old by ship standards. The current commanding officer (CO) had been an ensign on it twenty years earlier, and I had his old job as the cutter's first lieutenant in charge of the twenty-five-man deck force. That was not necessarily a good thing. But I was thankful I landed somewhere warm and where drugs were running.

The welcome aboard was not what I expected. The ship had been in a Houston shipyard for almost a month at that point. A hurricane was approaching, and two sailors had gone out for a party the night before and were in a horrific motorcycle accident. One was killed instantly, and the other was in intensive care and only lived another week.

As the boot junior officer, I was assigned numerous collateral duties that included decedent affairs. My job was to inventory all their personal effects and have some contact with the family. As a twenty-two-year-old, this was real, traumatic, and not at all what I envisioned. It gave me insight on how to engage someone with whom I had no previous relationship and deliver difficult news. The executive officer (XO) knew I was, perhaps, more prepared than the others with the recent loss of my friend Peter and my granddad. By volunteering for this very unpleasant but necessary assignment, I was given the most desired position on the ship for my rank—the weapons officer. It lined me up for leading the law enforcement program and boarding teams.

After a few weeks, I was off to the Maritime Law Enforcement School in Yorktown, Virginia, which I very much enjoyed. Upon graduation, I flew to Guantanamo Bay, Cuba, to meet the *Valiant*, which was undergoing refresher training following the shipyard. To be clear, this was naval warfare training and was perhaps the most stressful part of any ship tour, Navy or Coast Guard.

God help the nation if we had to use the *Valiant* in a modern naval battle. While the Navy ships had automated advanced weapons systems with missiles and torpedoes, we had a three-inch gun with two crank wheels operated by two guys to get on target . . . hopefully. It was right out of a WWI movie. But I had two great gunners' mates who kicked butt.

Then it was time for me to show my skills on the bridge. The Navy and Coast Guard exercise evaluators were quick to kill off anyone who looked like they knew what they were doing. Meaning, as a brand-new ensign, I was the one guy they spared in the final battle problem.

The bottom line was we made it through the training successfully—no thanks to me—and then we were back to our business of catching drug smugglers and interdicting illegal migrants attempting to sneak into the country.

The only person on board more anxious to get a drug bust than me was the CO. He was an old-school leader and a stern CO. I would classify his effective leadership style as fear based, which he excelled at. This style was actually very common in those days among ship captains. He had no problem with getting very upset and grilling you with colorful language in front of others. Being brand new to the fleet, my skin was still pretty thin. Those occasional, spirited, counseling sessions caused me to panic and, at times, almost lock up, which if navigating in confined waters, could be dangerous. His leadership style is excellent for fighting watchstander complacency but also led to the negative consequence of his crew avoiding or delaying what could be seen as bad news.

You can always learn leadership lessons, both good and bad, from almost any boss, especially in uncomfortable situations. The key is to learn from it, pick those things that will work for you and then apply in the future when you have your own command.

In my opinion, the XO was a great leader who genuinely went out of his way to develop the officers and crew. He was the insulation I needed during this initial tour. The combination and contrast of the two shaped my future command style.

First ship assignments were only two-year tours. As the class ahead of us rotated off the ship, I was now considered a seasoned ensign and organizationally promoted to department head. I was lucky to land the first lieutenant job, meaning I was responsible for the twenty-five-person deck force. Now, nearly twenty-four years old, I was a supervisor of a new ensign, a chief petty officer, other petty officers, and a whole bunch of seamen. Some of these guys already had twenty years' experience, so I'm not sure who was supervising whom.

We even had a seaman who had been a bank manager. Despite managing more employees at the bank than we had in the department, he declined the offer to take on more leadership responsibility. He was a bright, hardworking guy but said he was fine where he was. He told me the reason he had left the bank was for a hands-on adventure and that the significant pay cut was worth it.

I had another seaman who was the one guy we hid whenever there was an important official visit. He was a high maintenance employee, as the chief would say. Ironically, his parents were millionaires, and they had made a deal with him. If he completed a tour in the military, he would inherit one million dollars. To my knowledge, he still isn't a millionaire.

One thing that helped me as a junior officer was that I got along well with almost everyone, and more importantly, I was willing to listen and learn. Another thing was that I was gung-ho about the missions. When I had duty as underway officer of the deck (OOD), I had a standing agreement with the bridge team. If anyone on the bridge spotted a drug boat, I bought him a steak dinner. (If you are wondering, *Valiant* had an all-male crew, so the offer didn't appear creepy.)

It had been a dry first year for noteworthy operations, but things began to change my second year.

Valiant departed Galveston for Hampton Roads, Virginia, for two weeks of intensive training to polish their firefighting, damage control, signaling, communications and navigation skills. Hampton Roads was home to the world's largest naval base that included 75 ships, 134 aircraft and 82,000 sailors. While on the bridge, we navigated into this incredibly complex port with a cruise terminal, trains, trucks and towering gantry cranes handling 1000-foot-long container ships along with mountains of coal to be shipped overseas. Then when we made the turn to the south down the Elizabeth River, we passed huge aircraft carriers which dwarfed the tiny *Valiant*. In Boy Scouts ten years earlier, I was very nervous driving and mooring a fifteen-foot motorboat with its complex mix of steering wheel, throttle and ropes, not to mention the wind along the pier. But I earned that merit badge. Now I realized my initial impression of *Valiant* being overwhelmingly complex was a perception of a naïve twenty-two-year-old. Getting out of your comfort zone and seeing the real world has a way of building personal confidence and understanding.

Our next mission included conducting training and then a search and rescue exercise in the Bahamas with international partners. Sure, we had received theoretical training on this type of mission at the Academy, but it would be another thing to use a helicopter and small boats to simulate it.

Before arriving in Freeport, *Valiant* diverted to intercept a thirty-eight-foot sailboat off Nassau with 102 Haitian refugees on board that was sinking. This was not an exercise.

With no lights and a wooden hull that didn't reflect radar well, the boat was located better by smell than sight. These were horrific conditions for desperate people. It sounds impossible that 102 people could be on such a small vessel, but it's true. Men, women and children packed themselves in every corner

Sailboat with Haitian refugees
(photo courtesy of Coast Guard DVIDS)

with barely room to move—meaning minimal space for food and water. And no bathroom accommodations.

There had been fatalities among the Haitians before our arrival. Rather than keep the deceased aboard, they had thrown them over the side. Upon coming alongside, the Haitians anxiously reported through our Creole interpreter that a tiger shark had been following them for days. With the boat sinking, they were eager to get off.

Approaching a very rickety boat at night takes coordination. The

Coast Guard had previous cases where those onboard panicked and rushed toward the cutter's small boat, capsizing their vessel. Can you imagine over one hundred people in the water without life jackets at night, many who cannot swim? Therefore, we did not approach such a vessel at the beam. It's much better to come from the stern and pass life jackets before taking anyone off. Our crew members, many of them parents, showed great care for children and even pets. This, in turn, kept the adults onboard calm.

I was the boarding officer for this case as we methodically removed the migrants from the vessel. Normally we would remove those with medical issues, women and children first, but they were so tightly packed, we just took who was closest. While this was taking place, the deck force aboard *Valiant* erected a tent structure on the flight deck to house, feed and provide medical care for the migrants.

We completed the transfer operation by 0200. All 102 were safely moved to the *Valiant's* flight deck. I then boarded the sailboat to ensure everyone had been removed before we sank it as a hazard to navigation. *Valiant* added twelve more migrants from a second cutter bringing the total to 114 that would be repatriated in Port-au-Prince, Haiti.

Valiant's Flight deck ful of Haitian migrants.

Haiti was rapidly becoming a failed state, meaning anyone in that country with the means and ability took to the water to escape. This was especially true when our national leadership made comments that could be interpreted as migrants could remain in the United States if they made the dangerous journey. For the United States, this represented a huge threat to life at sea; therefore, the Coast Guard always stationed cutters in the vicinity of Haiti for both interdiction and rescue missions.

While Coasties certainly made less money than many not serving in the military, few careers could rival a story where their actions no doubt saved numerous lives in dire straits on a sinking boat surrounded by sharks.

After dropping off the migrants and a brief fuel stop in Guantanamo Bay, Cuba, *Valiant* was back underway in the Windward Passage between Cuba and Haiti. We came across the fishing vessel *Holy Mercy* that was riding low in the water. Several indicators, including the crew's behavior and the comment that the captain was no longer aboard, meant a high probability it was a smuggling vessel. The captain of a smuggling vessel often received harsher sentencing, so not fessing up to being a captain was a common practice by smugglers caught in the act. As the boarding officer, I went through the routine of initial questions and then conducted a tactical sweep of the vessel for officer safety.

As soon as I entered the cabin, the smell of marijuana hit me. The boat was loaded, and the crew knew the jig was up. I cannot express to you how great a feeling it is—or perhaps relief—to call back to the *Valiant* and let them know we found drugs on board! We were due for a bust, and we were stopping five tons of drugs from entering the United States. All the boardings, drills, and bouncing around seasick just paid off. We had discovered 200 bales, approximately fifty pounds each, of marijuana in the fish hold. Three crew members were arrested. The vessel and contraband were seized, having a street value of approximately 24-million dollars. The euphoria eventually gave way to a ton of paperwork on a smelly, bobbing boat and ultimately testifying. But it's absolutely worth it!

Later, as witnesses for this case, Petty Officer O'Rouke and I flew to Miami to testify in the prosecution of the three smugglers. The government funded us one hotel room, and we met with the district attorney before taking the stand to ensure we knew when and

Jim Loew watching bales of marijuana come aboard the cutter.

where we needed to report. He also discussed questions we might get from the prosecutor and the cross-examination. When we returned to the hotel room, there on the threshold was a live bullet. We admittedly got a little freaked out and practiced escape routes from the hotel. We also advised the district attorney, but he said it was normal. It was, after all, Miami in the 1980s. I will say this was a lesson learned on staying alert to my surroundings, which set me up well for future high-risk missions overseas.

The bottom line, the crew of the *Holy Mercy* received little mercy. They were convicted.

But back to the patrol. After disposing of the *Holy Mercy,* prisoners and evidence to the court system, *Valiant* headed north for what was expected to be two days of relaxation in Nassau. The day before arriving, *Valiant* was once again diverted. This time it was for a converted shrimper with Haitians being detained by our partners in the Bahamian Defense Force.

When we boarded the *Kerleen Deborah Express,* we quickly detected explosive gas in the vessel's bilge. In an abundance of caution, we removed the Haitian refugees as quickly as possible to *Valiant* and then ventilated the vessel. But there was something else very odd about the crew. It was composed of fifteen women and one man claiming to be the master. It was extremely unusual, and despite searching the boat, we could not find anyone else.

To be clear, this was not Donald Trump's yacht! There were even goats onboard, so you had to watch where you stepped. I wasn't convinced we had everyone, so I pointed to *Valiant's* mounted .50 cal machine gun. Through the interpreter, I told them we had to sink their vessel as a hazard to navigation and pleaded with them to show us where anyone else might be hiding. The bluff worked. They showed us where men were concealed in the bulkhead in the front part of the bridge and others submerged in a tank. I suspect the original fumes were coming from the open tanks. After feeling confident that all people were accounted for, thirty-eight were transferred to the *Valiant*. Another boarding team member and I remained on board to pilot the vessel to Great Inagua Island in the Bahamas.

The vessel looked like it was built a thousand years ago. Voodoo bones and chicken parts adorned the bridge, and a goat roamed loose. The smell was a combination of fish, smoke, gas and a port-a-potty. I had only a handheld compass, binoculars and a tiller for steering. Below the deck was not much better than above. The wood planks of the hull showed obvious signs of rotting, no doubt exasperated by the constant water in the bilge. The previous occupants coached us on the use of a rubber boot on the end of a pole contraption to dewater the bilge. To add to our anxiety over our unseaworthy journey, at one point, a waterspout—a tornado at sea—passed between us and the escorting *Valiant.*

I bumped bottom once going into Great Inagua, but eventually, we turned the vessel over to the Bahamian authorities and hopped a small boat ride back to *Valiant.* I felt incredibly sad for these folks being returned to Haiti, the very country they desperately fled from. But if we didn't follow the deterrence policy, even more migrants would be emboldened to make these perilous journeys.

A few months later, *Valiant* was underway again for a forty-two-day patrol. After a brief stop in Miami, we located three bales of marijuana floating off Florida's coast. Technically it was considered a seizure since someone could come along and pick it up and distribute it ashore. But we didn't have any arrests with it. Because dead chemical lights were attached to the bales, we suspected they had been dropped from a smuggling aircraft. We felt we were getting closer to another good bust, which would include catching smugglers in the act.

Bust or not, after being pent up on the ship for several weeks, it was always exciting to get off and explore a new port. Let me be frank. Back in those days, when I say explore, I mean drink alcohol. One of our first stops was the famous bar on Duvall Street in downtown Key West known as the Bull.

Much to the owner's delight, a large contingent of *Valiant's* crew had already set up shop. One of the engineering petty officers who was already playing pool asked me to team up with him and play a couple of British sailors who were also in port off the HMS *Black Rover.* My teammate sweetened the deal by saying the losers would buy the winners the next round. They agreed and it was on. I'm not sure my play helped the team that much but we did manage to edge out a victory. When it was time to pony up drinks the already drunken British sailors reneged on their responsibilities. It didn't bother me as I had linked up with another junior officer from *Valiant* who eagerly wanted to cross the street and check out perhaps the most famous bar in Key West, Sloppy Joes.

A handful of us made our way over to check the place out. Out of nowhere, my pool teammate, sporting a bloody nose, sprinted into Sloppy Joes with a gaggle of British Navy guys in hot pursuit. I jumped between him and the pursuing mob and tried to diffuse the international misunderstanding, diplomatically. Before I knew it, I got sucker-punched in the face. I jumped the guy. That started a crew-on-crew mini-battle that was not appreciated by the staff of Sloppy Joes.

While the bartender did not see the first punch that I "valiantly" blocked with my face, he did see me jump on the guy, so I got evicted until other patrons told him the real story, and I was allowed to come back in. Skirmishes between the British Navy and *Valiant's* crew continued throughout the evening. Eventually I guess the *Valiant's* crew were not that eager to continue the brawl so they would actually turn on each other, British Navy versus British Navy. Sailors don't get paid much, and being couped up on a ship for months at a time will often mean time ashore is a priceless opportunity for adventure. They do tend to see a lot more of the world than most people and understand how precious free time can be.

When I got back to the brow of *Valiant*, Jim Loew, the XO, met us. He looked my bleeding mouth. "Did you have a good time?"

"Yes, sir!" I guess other folks had come back and passed along our reenactment of the Revolutionary War.

After two days in Key West, I'm pretty sure the XO wanted to get the ship underway. *Valiant* headed south to rendezvous with the Honduran Navy for joint law enforcement operations, marking my first international engagement. Upon arrival in the operating area, *Valiant* welcomed a Honduran officer on board while providing Coast Guard representatives to Honduran gunboats.

In addition to law enforcement activities, *Valiant* responded to assist a sinking Honduran fishing vessel. Efforts by damage control personnel lasted all day and salvaged the vessel, rescuing its thirteen crewmembers.

We located and seized the fishing vessel *Kimberly*, a sixty-foot US shrimper headed northbound off the coast of Nicaragua. The vessel had three US citizens and a huge load—600 bales of marijuana, estimated to be 33,000 pounds. Three smugglers were arrested, and a custody crew was placed on board for the transit north. However, after a day of travel, the *Kimberly* began sinking. It was a scramble, but sixty-three bales of marijuana were taken off as representative evidence, and the custody crew was removed within an hour.

It was a good run on *Valiant*, and I boldly requested to be screened for my first command, which would have been an eighty-two-foot patrol boat. My only bad mark had been in the first year for becoming too familiar with the crew.

I screened for command, but few boats were available, so the best alternative was to be an XO on a new high-speed 110-foot patrol boat. Jim Loew encouraged me to jump on it and, based on my counter-drug passion, recommended that I pursue Patrol Boat Squadron Two based in Roosevelt Roads, Puerto Rico. Sure enough, it was great advice, and I landed the job on the Cutter *Vashon*. Shortly after receiving my orders, the command was given to paint the entire squadron gray, load everything on transport ships, and move to the Persian Gulf as tensions were rising in the Middle East.

Thankfully, the Bush administration rethought sending these particular boats since they were seen as the tip of the spear in the drug war. I got to stay in the western hemisphere, at least for this tour.

Lessons Learned:

Change your car's oil regularly; you are not a college student anymore

Be tenacious, patient, and positive. It will eventually happen.

Volunteer for dirty and unpleasant jobs.

Job satisfaction is not always about the money.

When dealing with desperate people, show great care, especially for children and pets

Do not let panic paralyze you. Stay laser-focused on completing the mission

Sometimes a bluff of uncovering a gun mount works wonders.

Chapter 4

Front Line of the Drug War in Puerto Rico

THE BEST PART ABOUT making lieutenant junior grade is that you're no longer an ensign. You have ever-so-slightly more respect but can still occasionally make mistakes and get away with it.

The challenge with moving to Puerto Rico is that all household possessions, including a car, must be shipped by barge. My Camaro had held up fine during spring break and in Texas. Then she was on a barge bound for San Juan.

With a little R & R coming to me, I decided to make a solo Space-A trip to see how far I could go west in a couple of weeks. Sure enough, I made it to Sydney, Australia, and back, meeting new folks along the way. I hiked in Australia's Blue Mountains and up Diamond Head in Hawaii.

Even with this trip in the middle, I beat my car to Puerto Rico by a month, so I relied on shipmates and a bike as a means of transport.

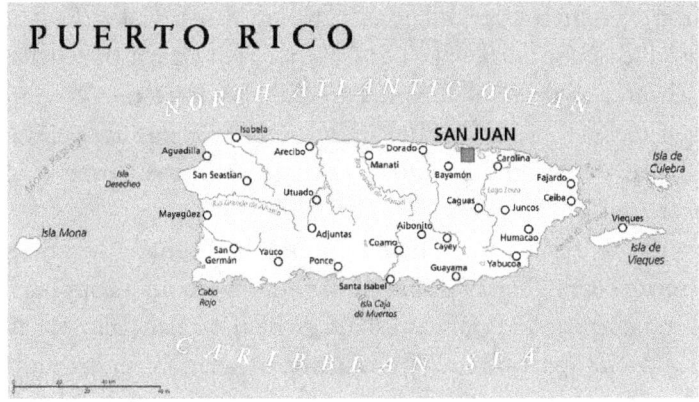

Patrol Boat Squadron Two tied up at the end of a long gravel road and consisted of a small building that housed a squadron commander with an eye for details, spare parts and administrative personnel. I suspect it was similar to a temporary World War II patrol torpedo boat base tucked away in the jungle. The single pier was long enough to tie up a 110-patrol boat on each side. The squadron consisted of five boats with a sixth boat in San Juan that occasionally joined us. If more than two boats were in at the same time, we had to breast out, meaning tie up to another patrol boat instead of the pier. When the boat next to the pier had to leave for patrol, we had to get the outer boats underway to accommodate their departure. It became known as the morning shuffle. Eventually, the boat COs let the XOs move the boats before they arrived. This became the first time I was the senior person on the ship underway, ultimately responsible for the vessel's safety.

That sounds pretty easy because we were effectively moving parking spaces, but things can go very wrong very quickly. On one such occasion, a patrol boat up ran up on the jetty, knocking over a bike that tumbled into a car. The whole squadron rallied to push the cutter off before the squadron command and other COs arrived to witness the spectacle. It took significant repairs but was eventually back in action.

I briefed this story thirty years later in one of my classes, and one of the students that came from that boat said the ship still pulled to the right.

J.J. Fisher was an Academy classmate that landed the sister XO job on the Cutter *Ocracoke*. We shared an apartment in Fajardo, twenty minutes north of the naval base at Roosevelt Roads on the east coast. Our apartment was on the twenty-third floor of Dos Marinas, two high rise towers that overlooked a lush rain forest called El Yunque to the west, and Palominos, Isla De Culebra (Island of the Snakes), and Vieques Island to the east. Further to the east were the Virgin Islands. On a clear day, we could see their faint outline. It was always warm, but up that high, we had the benefit of a nice easterly sea breeze and were rarely bothered by bugs.

Beyond the lack of bugs, being up high had other advantages. Early on, we received intelligence that smugglers were air dropping cocaine to small boats near Palominos Island, which was the closest island to Fajardo and our apartment. With our boss's permission in San Juan, we rented a sailboat and conducted a stakeout over a long weekend. There are specific rules against doing that now, but this was thirty years ago.

We were out on the right weekend, except the airdrop occurred further east near Culebra Island. I was less than satisfied to come up empty handed as we had asked for crew volunteers to give up their weekend for this mission.

A few weeks later, as it was getting dark, I was out on the apartment's balcony listening to Guns and Roses "Welcome to the Jungle" when, sure enough, a plane that met the description of the suspect aircraft flew overhead. A small boat emerged near Palominos Island and started flashing lights. It was going down perhaps a mile away. These guys were very bold! I immediately phoned the command center in San Juan, and since it was a non-secure line, I had to paint a story using code. They figured it out and launched the ready boat that happened to be my roommate's cutter, the *Ocracoke*.

As they departed Roosevelt Roads, they picked up an unlit contact heading south at a high rate of speed. This suspicious contact was between the southeast coast of Puerto Rico and the Cutter *Ocracoke*, which had also launched their small boat with a boarding team. Cut off, the suspect vessel increased speed, turned due west, and headed right at the coast. It was apparently like a scene out of a movie. The boat jumped Highway 3 and crashed into the woods on the other side. By luck, no cars were hit. Cocaine was found in the boat but no smugglers. It was believed later that the boat's occupants most likely jumped off just before hitting land and swam ashore. Our squadron got an assist for the bust, and J.J. and the crew of the *Ocracoke* had a very cool story to tell.

I was incredibly lucky to land on the Cutter *Vashon*. I had, in my opinion, the best CO in the squadron. The last thing I wanted was a live-aboard, single CO who was on their first command. Crew members—especially the XOs—would rather not have the big boss around 24-7. LT Mark Rutherford had already successfully commanded an eighty-two-foot patrol boat and was married with a couple of young kids. He was always calm and patient, and very funny with a quiet, sarcastic wit.

Our crew was also fantastic. Two-thirds were Puerto Rican, so my only complaint was, sometimes in the heat of battle—such as mooring the ship—they defaulted to Spanish. I really should have learned Spanish on that tour.

We all wanted the same thing—to lead the squadron in drug busts. This was not unlike trying to be the best patrol in a troop of Boy Scouts or the top company at the Academy. As the XO, not only did I have all the administrative tasks, but on a patrol boat, I stood multiple four-hour watches a day as officer of the deck, which essentially means driving the ship. I preferred the four to eight shift, because I got to see the sunrise and the sunset, which are spectacular in the Caribbean. I also served as the primary boarding officer and the away team leader. First Class Boatswains Mate (BM1) Hudson who used to act as boarding officer had been injured on a previous tour while boarding with the Navy and preferred to run the boat and the deck crane. That was fine with me!

On the larger *Valiant*, I had months of break-in duties and a formal board before I was deemed qualified to run the bridge watch. The bridge and combat information center or CIC had a team of people doing the navigation, shipping radar, radio, logs, the helm, lookout and engineering.

My first watch on the *Vashon*, I was blessed as qualified. On the 110-foot patrol boat, there was a lookout on the flying bridge and the OOD. That was pretty much it. There was an engineer, but they remained in the engine room for the most part. So, all those other tasks fell to me.

The first night, all the alarms kept going off on the bridge. Frankly, I felt a bit overwhelmed at first, but I settled in quickly, as did all the OODs. Unlike the *Valiant* with a top speed of eighteen knots down swell, the *Vashon* locked in at thirty-one knots and could chase down almost anyone. *Vashon's* hull, though, was like a beer can—built for speed, not for ramming. The other tricky thing was an eight-second delay on the throttles. If the throttle handles are in reverse coming in too fast to a pier, those eight seconds can feel like an eternity. The throttles of these A and B class 110s also clutched in at eleven knots. It was a rocket ship with lots of power to get you into trouble in tight marinas but also enough power to get you out of bad situations.

There was no wardroom for officers; everyone ate on the mess deck, which had an open galley. That meant we could forage for snacks at any time of day. The walls were adorned with as many weapons as there were crew . . . sixteen of us. We could make our own food, and we always had movies running. Back then, we had the new VHS tapes. There was no satellite TV on these early boats.

When our new cook reported, he stepped off the plane and complained of serious back pain. He was medically ruled out from sailing. That meant we were without an assigned cook for the next six months. A cook on a ship is effectively running a small restaurant. In specialty schools, they learn how to cook, balance menus, manage inventories, and process complex budget reports. Without a cook, we had to figure all that out on our own. The crew took turns shopping and cooking, and I did my best with the paperwork. About five months without having a cook onboard, we had the assignment officer for cooks down from Washington to meet with the squadron at all hands. We had been asking him about the status of the relief cook on a regular basis. We did it once again in front of the whole squadron.

He said, "I don't see what the problem is. Can't you cook hotdogs?"

It was good that I was unarmed. Within a month, our pushback worked, and we received a new permanent cook—this time with no back issues.

As you can imagine, people pay big bucks to fly to Puerto Rico and go marlin fishing. We did it for free. We had a walking fighting chair, a stash of poles, and spear guns. We were always on the lookout for debris in the water, and while we hoped to find a person in need of rescue or bales of drugs, a tree stump was almost as good. Out there, anything drifting had schools of fish around it—including marlin.

Mostly we caught bottom fish like red snappers or dolphins like this one, caught by Petty Officer Mark Ruble.

At least for the two squadrons based in Miami and Puerto Rico, the boat's reputation was generally built on the number of drug busts they made. We had marijuana and cocaine

Petty Officer Ruble was our best fishermen.

stickers on the side of the bridge wing, just like the *Valiant*. It was always a great day to put on the next sticker that meant we were earning our paycheck.

While we were part of a squadron, we had much less direction or connectivity from above as we do today. At best, radio communications and satellite navigation were sketchy in the Caribbean. There were occasions when we took the small boat into an island to make a landline phone call to the command center in San Juan.

Our schedule was rigorous by patrol boat standards. The average patrol boat in 1988 ran about 2,000 underway hours a year. When they were not at sea they were often in a ready response status meaning they had to get underway for missions if called in two hours. Most of the patrol boats in the Coast Guard were homeported by themselves and spread out along the coast for even coverage. But the two squadrons were different in that they had modest support staff of engineers ashore and by being clustered and a boat almost always underway, they didn't have to stand the ready recall status. This means when they were in they would not be called out. So they adjusted the squadron boat hours up to 3,000 hours a year. We had effectively ten days at sea a month with a break in the middle for food and fuel. After the patrol, we had roughly twenty days in port when we would rarely be recalled.

I started dating a Navy LTJG on the base who also resided in Dos Marinas. I found this schedule to be almost perfect for both of us. After being apart for ten days, it was great to get reacquainted.

We couldn't ask for a better patrol area either. There was a lighthouse at the southeast tip of Puerto Rico called Point Tuna, near Patillas on the previous map. We were told to patrol either east or west of Point Tuna and make drug busts. We could pretty much go anywhere looking for bad guys, and frankly, at the time, intelligence was quite poor. We were lucky if we could get a fixed-wing overflight to help locate possible targets to board.

To the west was the Dominican Republic; to the south, the ABC islands just off Venezuela: Aruba, Bonaire, and Curacao. We made it a point to try to hit almost every Caribbean Island. I think my favorites were St. Kitts, St. Maarten, and Barbados to the east, but all had their own charm.

For instance, Mona Island is uninhabited and lies in the passage between Puerto Rico and the Dominican Republic. This seven-mile-long, four-mile-wide, mountainous island was once mined by the Nazis during WWII for bat guano. It was honeycombed with caves and had a plateau on top with an airstrip. We sent landing parties ashore and hiked the area to check the airstrip for activity. The island was home to six-foot-long iguanas that looked more like dinosaurs than lizards. When a new crewman joined us, we had him lead the party up the trail to check the airstrip for activity. We didn't warn him of the island's true inhabitants until he stumbled upon one. That gave us a great laugh.

The squadron boats were tasked on occasion with rounding up migrants stranded on this waterless island by smugglers telling them they had reached Puerto Rico. We even discovered cocaine in the caves, no doubt waiting to be further transshipped. When the official checks were done, the deck force went out with spear guns to catch lobster, and we moved the grill to the island.

On one of these ten-day trips, Mark Rutherford's third child was due, and Mark let me take the cutter out as acting captain without him. He came to my stateroom, sweat on his brow and hands shaking as he passed his ceremonial command afloat pin to me. It was like the first time you drive the car solo, which is a bit scary. But I was not by myself; we had a great crew. On that trip, we pulled into Santa Domingo, Dominican Republic, needing to refuel. The truck driver refused to believe I was the acting captain as I looked young then. Those were the old days! Chief Carr, our engineer, handled it. I believe he was the oldest onboard, perhaps just in his early thirties.

Allowing the XO to be acting for a trip that included foreign ports of call was rare even back in those days, but this did wonders for my confidence and future command roles. I would do the same for my future patrol boat XOs, and they turned out well.

Everything was going well in Puerto Rico, and we had made a few large drug busts. A press release published in Roosevelt Roads *El Navegante* captured some of the efforts:

> On May 12, 1989, while *Vashon* was patrolling off the southwest coast of Puerto Rico, a helicopter from Air Station Borinquen located a suspicious-looking vessel forty miles off Cabo Rojo. A sighting report was passed to the *Vashon*, and we sprinted to intercept. Unable to establish radio communications, we launched our small boat to investigate. Alongside the fishing vessel *Melpo JR*, we asked the standard questions, ascertain its nationality, and then boarded to determine its cargo. Once aboard, we immediately detected the "smell of victory"–the strong odor of bulk marijuana.
>
> An inspection of the vessel's cargo hold confirmed that *Melpo JR* carried more than 100 of the classic burlap-wrapped bales of marijuana. Chemical tests confirmed the contents. *Vashon* subsequently seized the vessel and arrested

the seven-man crew, all from Colombia, and took the vessel and prisoners to Ponce for further processing by the US Customs Service. The bales, 131 in all, were offloaded at Ponce and weighed in at an even five tons of contraband.

Mark Ogle turning over nine tons of marijuana and seized boat to customs in San Juan.

We made a second, larger, marijuana seizure. Following a tense nighttime gun-to-gun moment between our small boat and the crew of the eight-six-foot S/V *Merjon*. I served as a prize crew on the trip back to San Juan and offloaded nine tons of marijuana. But it would be 19 September 1989 that changed the legacy of *Vashon* forever.

Lessons Learned:

When shipping a car overseas you'll probably be waiting for a
   month for it to arrive.

A bicycle for transportation is both good exercise and faster than
   walking,

Learn all the alarms on the ship's bridge before driving solo.

Make sure you always have a strong fishing pole onboard.

There are *really* big iguanas on Mona Island.

Tactic: Find a way to surveil smuggling operation without
   compromising the safety of those bravely providing
   information—you have a duty to take action against criminal
   activity.

Tactic: If the patrol boat squadron commander asks if your
   boarding team SWAT suits are legal, the correct answer is yes;
   these are "very functional coveralls."

Tactic: When bringing multiple detainees aboard from a drug bust,
   consider having them sit by themselves out of sight of the
   others for fifteen to twenty minutes. Vary the time. You're not
   doing an interrogation at this point, but they don't know that,
   and you will create distrust amongst them.

Chapter 5

President Bush Sent in Troops

HURRICANE HUGO HAD FORMED over the windward islands and grew rapidly to a category five storm, causing havoc throughout the eastern Caribbean. The island nation of Montserrat to our east reported 99% homeless with many fatalities. The track of the storm was bearing down on the US Virgin Islands and Puerto Rico.

We got the word that the entire squadron along with the 278-foot Cutter *Bear* would sortie south toward South America to avoid the storm and then follow in behind to assist in recovery efforts. The *Bear* would not only escort the squadron but also refuel the smaller cutters if needed.

As the storm impacted Puerto Rico, we had limited news reports on the status of our families and homes. Message traffic was ominous.

The storm's eye had passed over the Virgin Islands and ripped into the east coast of Puerto Rico. Communications with Naval Base Roosevelt Roads was completely knocked out, which foreshadowed big trouble. There was also a report that a small ferry boat was rolling down the street in my hometown of Fajardo. That left me questioning whether being oceanfront on the twenty-third floor with its magnificent view was such a great idea. Sure, it was well above the flood line, but the winds gusting over 140 knots were devastating and threatened to topple any manmade structure.

The task force of ships made it to Ponce on the central south coast of Puerto Rico. This town was mostly spared, probably due to the mountains to the north and east.

Eventually, *Vashon* refueled and received permission to run reconnaissance on the east coast and hopefully establish communications with the naval base.

Roosevelt Roads, once a lush green tropical paradise, was now eerily brown and still. There were no leaves on the trees, no lights, and no movement. It was like a nuclear bomb had hit it. There were sailboats and other small vessels littered in the mangroves. It smelled too. I suspect sewage treatment plants were unable to keep up. With our small boat in the water looking for debris ahead of the cutter, we proceeded cautiously and made it to our pier, but there was no power or water. Eventually, we were able to gravity-feed fuel for the cutter and keep the generator running. It took three weeks for power to be partially restored. The cutter was one of the few places in the region where we could cool off.

Immediately after tying up, I desperately wanted to leave the ship to see if my girlfriend was all right and whether my car and all my worldly possessions made it through the storm.

My Camaro, parked on a hill at the base, survived but had strange vegetation growing in the back seat. Tons of debris littered the roads. Looters and bandits were very prevalent at this time in Puerto Rico. Over the next months, I had numerous flat tires and learned to change them in less than five minutes. My head was on a swivel, and there were many places that I just didn't stop at red lights after dark.

The drive to my apartment following the storm was an adventure. I had to hike in the last half mile or so due to the road being blocked by downed trees and powerlines. In the parking lot below the towers, my roommate's truck had two cars stacked on top of it. Apparently, there had been mini tornados as the hurricane passed through. The towers were thankfully standing but had sustained heavy damage. The elevators were out, but tenants could use the stairs. Twenty-three stories are a lot of steps to climb, and my mind had a long time to wonder what I would ultimately find.

This reminded me of doing flight ops with the seabag at the Academy. I thought four stories were bad. Exhausted, I finally made it to our apartment. It was a mess. We had a wall blown out and debris everywhere, including a ceiling fan spun off its mount and jammed into the ceiling. JJ's. room was destroyed. When I opened the door to my bedroom, it was just like I had left it; everything was still in place and almost untouched, even on the shelves. I guess it was lucky I picked the side I did. I met J.J. on the way back and warned him what he would find.

My girlfriend, who resided on the twenty-sixth floor of the second tower in Dos Marinas, had been thankfully unharmed. We made plans to hump all the stuff down the stairs—including my girlfriend's. Did I mention she was on the twenty-six floor? My girlfriend's dad was a senior CIA officer at the time, so the guys on the ship kid me about making sure I was an incredibly supportive boyfriend (or else!). She secured a house, and we lived together until the Navy eventually moved dependents out of base housing. Then three, nomadic, single, patrol boat XOs were given a house to share.

None of that happened right away, though, because the squadron received dire orders that first night back. They read:

Get underway immediately and proceed at the best speed and surround the island of St. Croix.. Do not go ashore without permission.

It further explained that when the storm was approaching, all inmates had been released from the prison, fearing it would be destroyed. Police and National Guard had not only failed to report to work, but they were seen looting. Looting and gunfire were prevalent throughout the island. There was no direct communication either, just some ham radio operators getting broken reports. The governor of the Virgin Islands, who was on St Thomas, was downplaying any issues. Remember, these islands get much of their revenue from tourism, and a safe reputation is critical for their livelihood. St. Croix was also home to a major oil refinery on the south coast.

The Cutter *Ocracoke* was one of the first to arrive on the northside of the island near Christiansted. They personally observed a running gun battle but felt helpless given orders not to go in or intercede. A helicopter overflight reported "Send Troops" painted on the roof of one building.

In route, *Vashon's* mess deck was abuzz with loading magazines and prepping weapons as we knew this was about to get crazy. Then shocking news came in from the bridge: the three squadron boats traveling together had spotted a large, adrift buoy, which was a major hazard to shipping. One of the boats had to tow it back to the base.

Rutherford was the most junior skipper of the three, and often, those mundane tasks fall to the junior man's boat. Fortunately, the Cutter *Nunivak* took the responsibility, and we proceeded on. Once near St. Croix, we were all assigned an area of responsibility. *Ocracoke* was already off Christiansted in the north, and *Nantucket* was directed to the west off Frederiksted. Their immediate mission was to flag down anyone who could provide an update on status ashore. There was a partially standing pier, but our orders were explicit: "Under no circumstances are you to put anyone ashore."

On the *Vashon*, we got the crappy sector on the southwest corner of the island, and our mission was to show the flag and provide a calming presence. Let me tell you, our crew was anything but calm. In our sector, reefs extended several miles offshore. Those needing a "calming presence" were not getting it from us unless they had strong binoculars.

We overheard Nantucket's radio traffic to cutter Bear, a 278-foot, medium-endurance cutter, saying that they finally had been able to flag down someone on the pier. The person reported major looting and violence, and freshwater was running out. After overhearing the report, Rutherford pitched a plan to Cutter Bear, suggesting that we move north to back up *Nantucket* off Frederiksted. The *Bear* was so busy, they just welcomed the idea.

Bear oversaw the overall mission in the Virgin Islands. The officer running the show was incredibly calm and competent.

The radio communications were very structured with passwords and call signs. At some point, it was so crazy that we said, "Screw it!" and started calling each other by the ship's names. It simplified everything.

We moved up and anchored maybe a couple of miles south of Nantucket, which we affectionately called the *Clam Bucket*. We had our big eyes out, which were strong mounted binoculars, and scanned the city. Smoke rose from the center, and people with guns were everywhere. It looked like a scene from the movie *Zombieland*. Then things dramatically changed.

Two young men, who later turned out to be reporters vacationing from the Washington DC area, had taken it upon themselves to secretly enter the water and make an almost mile swim to the *Vashon*, trying to remain undetected from those onshore. We pulled the exhausted swimmers aboard. Peter Grant was a better swimmer and recovered a little faster. We got them water, and Peter removed a white plastic capsule that dangled from his neck.

Inside were the names of seventy-seven people who were in dire need of rescue. Peter was with a group barricaded on the southside, and there was another group of forty barricaded in town. At great personal risk, the reporters had smuggled the names out. The forty people downtown were in Liberty Hall. The building next to it was firebombed by looters the night before, and we could still see smoke rising from what remained. The looters had delivered a message to the inhabitants of Liberty Hall— either they get out of their way, or they would be firebombed at dark. The group was composed largely of stranded tourists who did not have weapons but had improvised acid bombs from a car battery. Knowing they could not hold off the looters for long, they were planning to sneak out just as it was getting dark and head to the RCA Building. It was already mid-afternoon.

They have a saying in the military: "Don't be the senior man with a secret." So, we immediately relayed the situation to the Cutter Bear, who relayed it to the District, Atlantic Area, and Headquarters. It went all the way to the White House pretty quickly. The word back was a little odd. It was: You are authorized to evacuate all US citizens.

As a US territory, except for a few foreign tourists, there were over 50,000 US citizens on St. Croix. We were just a patrol boat, but we got the gist of the order. We finally had the green light for rescuing those in dire need. While waiting, we had hashed out a plan.

I went with *Vashon's* BM1 Hudson and link up with LTJG Glenn Gebele, who was the XO of the Nantucket, along with one of their big, scary-looking seamen. We wore orange life jackets over our body armor so we could be distinguished from the crowd and offshore cutters. Hudson and I grabbed M-16s. I'm pretty sure, in retrospect, I wasn't qualified on it, but it looked intimidating, and that's what we were looking for. Since we didn't really know where we were going, Peter also donned body armor, and we took the small boats close to the beach.

The storm had washed out most of the sand, so we jumped into waist-deep water and waded ashore with M-16s over our heads. People with guns observed us.

A man yelled, "Why you need those guns, mon?"

After pointing out where to go from the boat, Peter returned to the *Vashon*. Things were getting quite tense on *Vashon*, and they were readying a back-up landing party should we get into trouble.

When we came over the rise to the southside's first location, we were met with cheers. That was a rush! These folks were a mix of tourists and residents desperate to get off the island and to safety. Their freshwater well was one of the last available water sources on the island, and they had received threats that the locals would retake their island and kill everyone. Along with shouted threats, war drums and gunfire sounded throughout the night.

But our mission from the White House was specific. We needed to get to those in Liberty Hall under direct threat of the fire-bombing first. We wanted to get to them before they tried to move to the RCA building.

After we quickly explained what we were doing, one gentleman handed us a handful of car keys. "Be my guest."

An older couple who lived on the island agreed to escort us through the backstreets to Liberty Hall. I told them we would try our best to return for them. We piled in the car with gun barrels hanging out the windows and proceeded cautiously to Liberty Hall. We saw a police car in the distance, and when the policemen saw us get out, they sped off. As the forty folks saw us coming up the hill, they cheered, ran out and hugged us.

I called the ships offshore. The plan was to escort the crowd to the one dock still standing and have *Nantucket* and *Vashon's* small boats ferry them to the ships. I told the group to get their passports, wallets, and medicine since we would be leaving in five minutes. They moved quickly, and we piled the older folks in the back of a big truck. The rest walked inside our four-person armed perimeter. There was shouting and cursing from those wondering the debris laden streets. The once inviting tourist town had been lined with shops, restaurants and small hotels now looked like a scene from a bombed out city during World War II. There was no power, running water or sewage.

At one point, a car pulled in front of the truck and blocked our passage. I tapped the barrel of my M-16 on the windshield, and the driver moved. I called the cutter for reinforcements to hold the pier head. That included Seaman Brandon Spies from the *Vashon*. Once we got everyone on the pier, there was some relief as we started loading and ferrying people to the offshore cutters. It was taking longer than I had hoped, as a few brought suitcases.

Feeling the operation was secure with reinforcements holding the pier head, the four initial members of the landing party agreed to make our way back to the initial group and fulfill our promise. We took one more look and hopped in the car, heading off just when gunfire erupted. A woman on the pier, startled, dropped her suitcase on Mark Ruble who was trying to hold the small boat alongside the damaged pier. He said it weighed more than his dolphin. The rescued tourists started moving much quicker then.

A kid with a gun came around the corner of the building where the gunfire had sounded. The kid said a Coast Guard guy was shot or something to that effect. Brandon thought it was us.

We heard the radio and thought it was the guys holding the pier.

Brandon tactically rounded the corner and saw that an owner with a shotgun had shot a looter at point-blank range in the chest, and first aid was impossible. Thankfully, the report of Coast Guardsman being shot was not true. The guys in the ships offshore were now getting anxious, as you can imagine. Brandon and his *Nantucket* partner reported what they had seen by their handheld radio and were directed to evacuate ASAP with the remaining tourists.

I believe the ships initially wanted the four of us to make it back to the pier for extraction, but gunfire in the area precluded it. Within a few minutes, the small boats finished the evacuation of the remaining tourists from Liberty Hall.

Meanwhile, we headed back for the group of twenty-seven tourists and residents we had met when we first came ashore. We had promised to return for them and had to hurry as it was getting dark.

When we got back to the beach where we first landed, most of the remaining tourists were ready to go and thankful we kept our promise to return. Mike Hudson and I made a quick sweep and then got word from the offshore cutters that a group of armed men were coming down the beach toward our position. One of the last tourists was an eighty-year-old man who was badly hurt by a flying glass and had a heart condition.

Glenn Gebele jumped in the car and headed south away from the threat, searching for a suitable extraction point by a small boat. Meanwhile, I went with the remaining two Coasties and tried to stall the advancing men. One had his hand behind his back, but when we leveled our guns at him, he turned back into the woods. Fortunately, the beach was on our left, and there was a water body on our right, so the three of us just needed to hold that strip of land.

Glenn had returned. He found a beach south of the city which had a sandy beach where we could wade the people to awaiting boats. *Nunivak* was back in time to launch their small boat and was vectored to Glenn's extraction point. We piled the remaining people in the cars. Counting the four remaining landing team members, there were twenty-nine of us. Peter Grant and his partner were already safe aboard *Vashon*. I rode on the trunk as a rear guard with Mike Hudson. Just as the sun was setting, we got to the beach and carried the elderly gentlemen through the surf to Nunivak's small boat. We piled in ourselves, knowing this was a mission to remember.

Once safely aboard with my M-16 stowed, I tried to lighten our new tourist guests' mood by asking who were Americans and who were Canadians. Then I told the Canadians I would have to take them back. They laughed; Canadians are always good sports.

The news of the events got attention. The following New York Times article painted a fairly good picture of what happened on the macro level.

> Bush Dispatches Troops to Island in Storm's Wake
> By Dennis Hevesi
> Sept. 21, 1989
> President Bush ordered more than 1,000 military police
> to the Virgin Islands last night to help restore calm after two
> days of looting and sporadic shooting on St. Croix that was
> set off by the devastation of a major hurricane.

The article detailed firsthand accounts that witnessed looting by both police and national guard. While the Governor played down the violence on this tourist hot spot, the personal accounts of survivors told a very different story. US Marshalls were also deployed to recapture inmates who had been released before the storms arrival.

Unofficial reports said that 97 percent of the buildings on St. Croix, which has a population of 53,000, were damaged or destroyed by the hurricane.

In the aftermath of the mission, the *Vashon* received some nice letters.

In a letter dated September 27, Secretary of Transportation Skinner's letter to the 7th District Commander, Rear Admiral Martin Daniell stated that "the performance of personnel involved in the restoration of law and order on St. Croix is particularly commendable. Although the landing parties' lives were in danger, these personnel rapidly assessed the volatile situation on the island. It immediately provided a calming influence to the local residents. The evacuation of personnel who felt threatened was executed flawlessly by numerous Coast Guard vessels and aircraft in the area."

Mr. Edward Fink wrote Mark Rutherford a letter:

> Dear sir:
> Please accept my personal thanks for the *Vashon's* efforts
> made by you and your crew during the recent trouble in St.
> Croix. I was one of the first to be evacuated from the hotel
> in Frederiksted after two fellows staying at our hotel swam
> to your vessel. It was a marvelous sight to see your people
> come up the road with arms to escort us to your ship. My
> deep thanks to you and the entire crew for getting us off the
> island and taking care of us while on your ship.

Mr. Albert Morris wrote to Senator Packwood and Congressman De La Garza. "Were it not for the Coast Guard and their assistance, there is little doubt that I would be alive today."

And finally, perhaps one of our favorite letters was from Joyce Adams:

> To everyone aboard the *Vashon*. You rescued my daughter and me on September 20 off St. Croix. I just want to give you my wholehearted thanks formally. We had just experienced total anarchy on St. Croix for three days and had been caught inside gunfire in Frederiksted. I wish I could explain in words what I felt like when you reached out your hand and took us aboard. My daughter and I will never forget it. We are so grateful. Your job is so important, and you did it so graciously and so warmly. You made me a real patriot. Merry Christmas to you all!

The correspondence created interest in retelling the story by an author already writing a history book on the Coast Guard. Samuel Schreiner Jr and his wife visited Patrol Boat Squadron Two in Puerto Rico and conducted interviews. They also spoke to several evacuated tourists back in the United States. For whatever reason, he fixated on the *Vashon*. Within a year, he published his book: *Mayday Mayday! The Most Exciting Missions of Rescue, Interdiction, and Combat in the 200 Year Annals of the US Coast Guard.*

I sent this book to my father, who was stricken with cancer. He had gone downhill fast, from 200 pounds to about 130, and had little to smile about. Not much gave him happiness those days, but that book sure did. Not only was *Vashon* featured in its own chapter, but several of our drug busts also made it into another chapter. That is saying something, because the Coast Guard, over its two-hundred-year history of fighting in every war, had some incredible stories.

But we had less than twelve hours of downtime before we had to spring into action again.

Lessons Learned:

If you are going to date someone in a high rise, make sure they live on the first floor.

Have a jack in the car and know how to change a tire fast.

When evacuating civilians in a hurry, ensure they have passports, wallets and medicine.

Law and order are fragile after a storm.

Don't assume everyone in uniform is a good guy.

If you go into a gunfight, take the biggest one you can find

Thirty five years after this St. Croix mission, Former Attorney General Bill Barr published his best-selling memoir *Just Another Damn Thing*. In his book, he described his role in the Bush administration and the middle-of-the-night guidance he provided to the White House and responding Coast Guardsmen.

Chapter 6

No Time to Celebrate

THE DAY AFTER THE St. Croix rescue and delivery of the tourists to the Cutter Bear, *Vashon* proceeded to Charlotte Amalie in St Thomas for fuel and a little, short-lived downtime. The cutter received a call that there was a hijacking of a vessel in progress, and the Puerto Rican Police had their hands full and requested assistance.

We went from cold plant, which means all equipment was shut down, to underway in five minutes—the fastest we ever did it. Coming in quick with overwhelming force generally resolves most situations. It was over in minutes, no weapons discharged, and the cops took it from there.

After retaking the hijacked boat, I could safely say we were invincible as a crew; but the smugglers didn't get that memo. In fact, with all the destruction and disruptions, they saw it as a great opportunity to push through as much product as possible while we were distracted.

SN Schmidt and GM2 Huffman prepare to retake the hijacked vessel.

On October 3, 1989, while we were operating to the east of Point Tuna just south of Vieques Island, there was a report of a possible air drop. On our way back to homeport, we saw an eighteen-foot vessel heading west at a high rate of speed. Although we were heading home, we decided to check it out. One of the tactics we liked to use with the 110-foot cutter was to intercept at top speed to see how the other vessel would respond.

The rogue vessel was running just south of Vieques Island on a westerly course toward Puerto Rico. When we got close enough, we engaged the blue light and siren and attempted to hail the vessel with no response. Knowing the two men onboard were up to no good, we quickly suited up for boarding.

Tom Huffman, our gunner's mate, uncovered the bridge wing fifty-caliber machine gun. There were no bullets in it, but the other crew didn't know that. When they looked back, I pointed to the gun. Upon seeing it, they made a sharp turn due north and completely beached their boat.

If you have seen the Clint Eastwood movie *Heartbreak Ridge*, this was almost the exact beach location where they filmed the landing on "Grenada."

We had to rapidly get our small boat in the water because you do not beach a cutter if you want to keep your job. As I jumped in our boat, two men on their beached vessel grabbed a duffle bag and raced into the woods heading up the mountain. Their small boat was high and dry with their outboard still running.

The Vashon with its small boat, nicknamed Piglet.

After our adventures, we had nicknamed the *Vashon* the "War Pig," and her small boat was "Piglet." At one of the worst possible times, Piglet's outboard decided not to start. With adrenaline pumping, we used the boat's paddle and butts of long guns to get to the beach while devising our pursuit plan. Eventually, Piglet came to life, but only after the two suspects had disappeared with a ten-to-fifteen-minute lead. Two of my shipmates secured the engine on the beached boat so it wouldn't seize and did a quick check during which they found loose kilos under a compartment of the center console. Bingo, we had a drug bust. I pulled my 9mm and started solo after them up a creek bed. There were thorns everywhere, like inch-long thorns, so I knew the smugglers were paying the price as they were in t-shirts and shorts.

495 lbs of cocaine recovered after hot pursuit up the mountain Vieques Island

This was probably the scariest moment in my career. There were two of them in the woods, running up the mountain—which was incidentally a test range for the US Navy, so there was a possibility of unexploded munitions around. There were spiders, and I don't like spiders.

These two were desperate to escape; they had the high ground and who-knows-what in the duffle bag. Over the radio, I told the guys with the boat to stay together and circle up, perhaps finding a better trail to cut off their escape. We were using paintball tactics as we had no formal military training for this. The guys back on *Vashon* had gotten the authorities on the island spooled up. The cavalry arrived in the form of a FURA Puerto Rican Police Helicopter and a Navy K9 unit.

While we didn't find the suspects that day, we eventually found the empty duffle bag with a name stenciled on it, and then an ice bag in their cooler on the boat identified what village they came from. Eventually, our Customs and Drug Enforcement Agency (DEA) friends captured them.

We pushed the boat back into the water and drove it to our pier in Roosevelt Roads where we transferred 495 pounds of cocaine. But we weren't done. We got two more cocaine busts on a four-day trip.

On December 18, *Vashon* diverted for a possible suspected drug offload south of Santa Domingo, Dominican Republic. After requesting a C-130 overflight, they located the Coastal Freighter *Vita Nova* and vectored in the *Vashon* to intercept at dusk. The aircraft searched for the better part of the day before finding the suspect, but the plane had to depart, being low on fuel.

Once on the scene, we used our VHF FM radio, loudhailer, blue lights and siren to get a response. Then we launched our small boat to communicate alongside. *Vita Nova's* response was silence, and they shut off their lights. They steered erratic course changes to the east and then the west. *Vita Nova* was larger than *Vashon*, but not as fast or nimble, so we stood clear. In the small boat, we also turned off all lights, took up station behind the vessel, and observed crew members on the stern throwing something in the water. We recovered torn papers and a ripped Colombian Passport. No flag was visible, and the only indication of a homeport was Valletta painted beneath the name on the stern.

At 1915 hours, we observed smoke coming from the interior starboard side of the ship. The vessel started to ride lower in the water. The vessel slowed and eventually stopped, indicating it was rapidly taking on water. The crew then appeared on deck and moved to the forecastle or bow, where they began the long vigil, waiting for their scuttled vessel to sink before disembarking.

On *Vashon* and alongside in Piglet, we made ready for an extremely dangerous rescue. The large vessel became unstable and could possibly capsize, pulling debris and the crew down with it. From the small boat, we requested names so we could inform their next of kin. We told them to write their blood types on their arms. We also discussed things like sharks. All requests were rebuffed.

We received explicit directions not to board from the District Command Center in Miami for either search and rescue or law enforcement unless invited by the crew, or if we had a statement of no objection from the flag state, or the crew abandoned ship. The sinking crew told our small boat that the master had previously abandoned ship, which was a common tactic by smugglers as the captain always received the harshest sentence. The spokesman from the crew said they would not allow the Coast Guard to board without permission from the company's owner in Barranquilla, Colombia. He also stated that the vessel was registered in Malta, the last port of call was Curacao, and they were heading to Haiti with a cargo of cement.

As the ship got lower and lower, the crew put on life jackets. At 2217 hours, it was apparent the ship could not be saved, and the crew of seven abandoned the ship to their life raft but refused any assistance from our small boat.

Once they were clear of the sinking freighter, we approached the raft with weapons drawn. This is one of those moments you remember: very dark, choppy water on a chilly December night. We towed the raft quickly to the *Vashon*, where our crew searched and detained the survivors. Then with the small boat, we returned to the sinking freighter now abandoned. The freighter had settled, and the cargo hatch in front of the pilothouse was awash.

We then made the decision to rapidly board the sinking vessel by coming alongside with the small boat. We timed the jump to get over the rail and not be knocked to our feet by the water on the deck that was washing back and forth as the ship settled abeam to the seas. In retrospect, this was probably very unsafe, but we yelled to make sure everyone was off. I had a sinking feeling that these smugglers would get away with it, so we decided to rapidly cut as many cargo straps as possible in hopes the ship would sink rapidly and possibly discharge cargo. Then it was a stand-off.

The seven Colombians were restrained on the bridge wing for officer safety, hoping the vessel would carry down any smuggling evidence while we were hoping for the opposite. One of our crew members, a young fireman, was Puerto Rican, but he looked like he had come down from the Scottish Highlands with light skin, red hair and freckles. We always put him on prisoner duty to stand there and listen. Overheard conversations can be extremely helpful in follow-on trials. But without drugs, we had no case.

At 0158, the vessel eventually sank creating a debris field. We began to sort through what we found on the surface. At 0206, we recovered the first of several duffle bags full of bricks of cocaine. The total weight recovered was 735 pounds, enough to convict. Later, when I took the witness stand, the defense made an argument that perhaps a submarine had discharged the cocaine. The jury didn't buy it.

Mark Ogle sitting on 735- pounds of cocaine after a long night south of the Dominican Republic.

After we transferred the detainees and contraband in Mayaguez, Puerto Rico, we were rounding the south coast feeling victorious when we got word that an airdrop was going down near Ponce.

We shut off the lights and sped up to thirty knots. When we got close, we launched the small boat in complete darkness. We wanted to have the element of surprise. We had two small vessels in the vicinity of Isla Caja De Muertos Island near the southern port city on Ponce. As we closed the distance to the boats, we could make out bales in the water with green chemical lights attached.

Silently, we waited for a pick-up boat. A sailboat picked up one of the bales then threw it back. We jumped them and detained both boats, three Germans and a Canadian. We then picked up 520 pounds of cocaine floating nearby.

Due to the proximity, we landed on the island to ensure no one was hiding in the woods or waiting to retrieve any other bales.

The media was on us again because we had grabbed *Vita Nova's* load a day earlier. They asked if the Coast Guard captured the two boats we had

After recovering 520 pounds of cocaine, Chief Carr and our landing party prepare to search an island off Ponce.

brought in. I explained they were simply detained because they were in the area.

It was determined the boats were in the wrong place at the wrong time and were released.

That completed, we were ready to end our four-day patrol, that had us awake for most of the time. It was particularly cool that Mark Rutherford's little brother, who was then in the Navy, had been riding with us. Having the extra set of hands aboard was helpful, and he had some great stories to tell his Navy buddies.

On the way into the squadron pier, we blasted Eric Clapton's "Cocaine" on the loud hailer. It was a white Christmas for *Vashon* and a black month for the drug runners.

On April 4, 1990, I was subpoenaed along with Seaman Brandon Spies for a murder trial on St. Croix. Most of the testimony I had given in other cases to that point was for the prosecution. We were requested to testify on behalf of the defense this time by painting a chaotic picture of Frederiksted. It turned out that as a LTJG, just a year ahead of Glenn Gebele, I was the senior federal official on St. Croix when this shooting occurred. Brandon was a witness to the aftermath of the shooting, although he had not seen it directly. We made it to St. Croix and met with lawyers. The city's police chief was in the audience and was himself under indictment for having looted property in his home. After some deliberation, the defense attorney opted to not have the Coast Guard take the stand. The situation was chaotic but would not have justified the shooter's action. He was convicted.

After Hurricane Hugo, with limited power and water on base, the Navy moved many of the dependents back to the states and freed up government houses for active duty members. My girlfriend found her own place, and JJ and I moved in with another XO, which was nestled in a nice neighborhood among the COs and their young families, and Axe and Crash. These two, very large, petty officers were from JJ's boat, and they liked to play loud music and drink. Because we were about the same age and single guys, they invited us over for a housewarming party. A few hours later, with an empty bottle of peppermint schnapps, J.J. and I crawled out the back, into the woods and safely to our house to avoid the military police.

The next morning came early with a knock on the door. Mrs. Rutherford and Chief Carr's wife wanted to lodge an official complaint about the loud party thrown by our next-door neighbors. They had called the MPs. As XOs, we needed to handle it. As they peeled back the door, we were busted—not for the party, but for living like bachelors. Mrs. Rutherford grabbed some of my clothes strewn strategically on the floor and said she'd be right back.

When she returned, she had a clipboard. This was dangerous territory, especially when they looked in the cupboards and refrigerator, and at our priceless couch that we found near the dumpster. After thirty minutes of agony, they were ready to leave. She told me politely to follow her to her house to retrieve my clean laundry. When we pulled it out, there were numerous stripes on her dryer's inner drum. Apparently, I had a few blue ink pens in one of my pockets that did the damage.

She said, "Mark, if you ever want to land a wife—and this goes for you too, JJ—you better start working on this list. Elevate your game, gentlemen; you're not frat guys anymore."

I shared my experience with my skipper and Chief Carr.

"You're lucky it was only thirty minutes," Mark Rutherford said.

I must tell you now—women are my kryptonite. Whether it's a girlfriend, colleague, wife, daughter, mother-in-law, you name it, I'm helpless. They have managed to elevate my existence.

I took her list seriously and planned to buy a house at my next assignment and acquire new furniture. The women never found out the XOs were at the party (unless they read this book).

When it was time to move on, Mark Rutherford knew his crew had accomplished much. To put this in perspective from an award standpoint, I received an achievement medal for the rescue on St. Croix and a second achievement medal for two years of XO duties and the drug busts. That was considered significant in the day. I received many higher awards later in my career, but these are some of the most cherished ones—along with the memories.

I was extremely lucky to have been part of a great crew on the *Vashon*. I wouldn't have any on-site relief with the new captain replacing Mark Rutherford, but I had heard good things and vaguely remembered him from the Academy as an upperclassman. Fast forward thirty years, and I would be team-teaching with Vice Admiral Scott Buschman in the Sector Commander Course and always point out our common *Vashon* roots.

Lessons Learned:

Most people will back down if you are confident and have
overwhelming force.

Smugglers expect you to be lazy—don't be.

If you run up a mountain chasing drug smugglers, wear thick pants.

If you get bit by a poisonous spider, do not delay treatment.

Outboards tend to break at the most inopportune time.

When you're in your twenties, you think you're invincible.
Sometimes it's just luck.

If you're going to do laundry at your boss's house, make sure you
check all pockets for blue pens.

Either learn Spanish (preferred) or find a Puerto Rican crew
member who has pale skin, red hair, and freckles to stand silent
security watch over your detainees.

Chapter 7

Desert Storm

I RETURNED TO YORKTOWN'S Maritime Law Enforcement School, this time as an instructor. There was a familiar face in the school chief's office—Jim Loew. This was our third tour together if you counted the Academy. Shortly after reporting to Yorktown, war broke out in the Middle East as Iraq invaded Kuwait.

The two-day operation conducted by Iraq against its neighbor resulted in the country's seven-month Iraqi occupation. The invasion and Iraq's refusal to withdraw by a deadline mandated by the United Nations led to the United States' military intervention—the first Gulf War—which resulted in the Iraqi forces' expulsion from Kuwait. The Iraqis set 600 Kuwaiti oil wells on fire during their retreat.

Iraq accused Kuwait of stealing Iraqi petroleum through slant drilling. Some feel there were several reasons for the Iraqi move, including Iraq's inability to pay Kuwait more than $14 billion (USD) that Iraq had borrowed to finance the Iran-Iraq War. On top of that, Kuwaiti high petroleum production levels kept revenues down for Iraq.

The invasion started on August 2, 1990. Within two days, most Kuwait Armed Forces were either overrun by the Iraqi Republican Guard or fell back to neighboring Saudi Arabia and Bahrain. Iraq set up a puppet government known as the Republic of Kuwait to rule over Kuwait and then annex it. A few days later, Saddam Hussein announced Kuwait was the nineteenth province of Iraq.

From there, Saddam Hussein began massing forces for a possible push into Saudi Arabia. The US Navy geared up and requested boarding

Chapter 7

team training from the Coast Guard, because the Navy had been directed to establish an embargo.

Essentially, before making a push to expel the invaders, you want to starve them of resources for waging a defense. Think of it as a siege of a fort.

The request message came into Yorktown, and having been there only a week, I decided to march right into Jim Loew's office and request to be sent. He knew me, and I didn't even get a word out.

He looked up. "You just got here."

The request was denied. I gave it a day or two, and no one else stepped up, but once again, I was told to stand down. On the third time (and I would not have done this if it weren't Jim), I made my case, and finally, I was approved for the mission along with Chief Tim Cavanaugh. This was the first of many adventures with Tim. We were anxious to get in the fight. Neither he nor I were married, and we didn't have kids at the time, so we were the perfect candidates. We were also a good team. He was older and wiser, and I brought the energy.

The first brief we attended was in a large, classified auditorium at Norfolk Naval Station, Virginia. It was loaded with Navy brass, and I'm sure more than one of them wondered why the boys in blue were invited.

The embargo was just a small part of the overall brief discussing complex war plans, including constant refueling and re-arming of squadrons of planes aboard aircraft carriers. The seamen's entire careers had been training for possible war. Now it was time for them to put their money where their mouth was. There was not fear so much as a palpable excitement in the room. It was all very interesting yet foreign to me.

Finally, the brief shifted to discuss Maritime Interdiction Operations or MIO ops. That was our turn to talk about boarding team training. When the audience heard that it was Coast Guard policy to always board armed with a round in the chamber and safety off, I thought they were going to lose it.

One senior officer said, "Hold on right there. You're telling me we're going to put armed sailors with bullets in their weapons on foreign ships?"

This was a massive culture shift for most who had been accustomed to standoff weapons. I could understand their concern. They had nuclear missiles, torpedoes and Gatling guns, all requiring a lot of oversight, or really, really bad things could happen. But they also knew that the Coast

73

Guard had recent hand-to-hand experience in the drug war, and we were the ship boarding experts. They agreed to let Tim and I sail with the 2nd Fleet for exercises and the transit to the Mediterranean Sea.

While sailing with the 2nd fleet across the Atlantic Ocean, Chief Tim Cavanaugh and Mark Ogle trained boarding teams on twenty-one ships.

Our team was linked closely with the fleet JAG or legal officer and used the flagship USS *Mount Whitney* as our home base. From there, we worked feverishly to create a comprehensive videotape showing start to finish what an embargo boarding should look like. At the onset of the mission, I was summoned to the flag quarters of VADM Kalleres, commander of the 2nd Fleet. He and I watched some of their videotaped boarding drills conducted prior to our arrival, and he asked me to provide thoughts on how they could be improved.

A Navy vice admiral was asking a Coast Guard LTJG for his opinion. I must admit, I had not expected that. I think this might have been to see if Tim and I were for real. Unfortunately, during the exercise, the day before reporting, a sailor had crushed his leg between the small boat and a ship. They were using the captain's launch—a cabin-style, fiberglass, small boat called a gig—to transport the boarding team, as it was the only boat they had.

I pointed out that we had gone to rigid-hull, open, inflatable boats since we do so many boardings. I also mentioned we wore a much lighter, bulletproof vest. They were using heavy flak jackets that were very bulky climbing up a freeboard and offered little protection from bullets. He took some of these ideas for action, and he even joined us during some of the training debriefs.

On one occasion aboard a destroyer, I questioned the CO's choice to land and deliver his boarding team via helicopter vice a small boat.

He seemed irritated. "That's how we always do it."

"Would you be able to land the helicopter on an oil tanker?" I asked delicately.

At that point, he turned to the admiral. "This is crazy. How do you expect us to become highway patrolmen with one day of training?"

The VADM said, "We're going to war with the Navy we have. We can make some changes and do just-in-time training, but we will be arriving in a month. Get on board!"

It was certainly nice having the admiral along.

Our reception on most ships was incredibly positive, but it was a hectic month or so. Just when we thought we would catch a nap, someone woke Tim and me, letting us know there was a helicopter of opportunity for the next ship. We grabbed our things and rolled to the flight deck.

Sometimes we landed on a flight deck; often, we'd be hoisted on the cable, along with our gear. Occasionally they sent us by boat. But we made it through two battle groups. Since Navy folks might be reading this, here was our itinerary:

1. First Battle Group
2. USS *Caron*, USS *Vreeland*, USS *Leyte Gulf*, USS *Haws*, USS *Turner*, USS *Santa Barbara*, USS *Platte*, USS *Trippe*, USS *Vulcan*, USS *Mount Whitney*
3. Second Battle Group
4. USNS *Roosevelt Roads*, USS *America*, USS *Virginia*, USS *Normandy*, USS *Pratt*, USS *Prebble*, USS *Gallery*, USS *Halyburton*, USS *Kalamazoo*, USS *T. Roosevelt*

We started with an in-brief that included the CO, XO, OPS and potential boarding officer. Then we moved to the boarding team that ranged from twenty to forty sailors and/or Marines. Some ships had one hundred volunteers, and we were asked by their commands to pick the most capable ten.

We had to watch out for the sailors who came with a flashlight taped to their pistol and ruled out the over-anxious candidates. We ran through the training video, tested them with shoot/don't shoot scenarios, and walked through boarding procedures and contingencies.

The last ship we were on nearing Spain was the Aircraft Carrier USS *Theodore Roosevelt.*

Normally an aircraft carrier would never get involved with boardings. When we landed by helicopter, I was approached by the senior Marine onboard who oversaw security. I suspect being a Marine housed with 3,000 sailors could get old. He asked if we could provide the training to his guys on the mess deck. We were spent by then, but it was the last ship, and Tim and I said what the heck. Little did we know, it was filmed and went out on close circuit TV throughout the ship.

Finally, our part of the mission was complete, and we caught a Cod flight, which meant we got the catapult ride! That was pretty cool. They dumped us in Rota, Spain.

When we landed, tensions were extremely high as the deadline for Iraq's withdrawal was only days away. I was a bit disappointed we didn't sail all the way to the Persian Gulf, but it was probably a good thing. Due to the hostilities and the fluid schedule, our flight arrangements were canceled. We rented a car and drove across Europe, making it to Ramstein Air Force Base for a lift home.

This was the second time I had been to Ramstein—the first had been on the space-A trip with cadet classmates six years earlier. When we got to the air base, there were hundreds upon hundreds of troops lining the terminal waiting for transport east. Iraq had 900,000 troops, ranking as the fourth largest army in the world. They were battle-hardened after the Iran-Iraq War and had taken over Kuwait in just two days.

In the terminal, these brave, US armed forces service members—some of whom looked to be eighteen and nineteen years old—showed palpable fear for their next mission. Unlike protesters, they showed great courage, risking their lives for the freedom of others.

A few months later, these same young patriotic Americans won one of the most decisive wars in our history.

Lessons Learned:

In the military, run toward the sound of gunfire. Volunteer for dangerous missions; that's the job.

Flak vests are bulky and do not stop bullets.

A rigid-hull inflatable boat is better for boarding than a fiberglass or metal one.

Even outnumbered, you can win decisively with the right training, discipline and equipment.

You do not have to be old to be an expert.

If you are old, it's always good to ask the opinion of younger people.

Always have a partner that can cover your weak areas and vice versa.

You go to war with the navy you have; that doesn't mean you can't improve on the way there.

If you're stranded in Europe, rent a car and drive to an air force base.

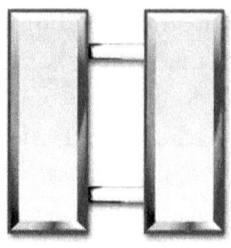

Chapter 8

Chief of the International Training Team

THE REALLY COOL THING about making lieutenant is the railroad track insignia. People will actually start paying attention to you.

You may wonder what happened to my Navy girlfriend from Puerto Rico. She was assigned to the Norfolk Naval Station, so we were within thirty minutes of each other while I was at the schoolhouse in Yorktown. At that point, we were growing apart. I had been deployed ironically with the Navy, and she had rekindled her relationship with an old high school boyfriend.

I had a brilliant idea, though. She had a yellow jeep wrangler, and I had a fiberglass camouflage canoe. I pitched the concept that we go to the Shenandoah River and take an overnight canoe trip for old time's sake. I had plenty of experience on river trips throughout North Carolina. She agreed, even if it was the last hurrah.

The Shenandoah River is in the northwestern part of Virginia, cutting through a pretty valley and eventually meeting the Potomac. The water was warm because it's very shallow and had numerous rock ledges. There were several rapids but mostly in the Class 1 and 2 range, with an occasional Class 3. I'd been in Class 4, so no worries.

We parked my car down the river, fifteen miles or so. When we finished the paddle trip, we would run back in my car to where we launched and retrieve her jeep. Everything was going as planned until about six miles into it; we started grounding the fiberglass canoe against the rocks. It was impossible to miss these scrapes, so we paddled to the middle of the river where it appeared to be the deepest. Once out in the middle, our equipment-laden canoe rapidly took on more and more water. I had a better vantage point of the situation by being in the rear.

I started bailing but couldn't keep up. Eventually, the water was at the canoe's edge.

I said, "Honey, it's time to get out."

I saw the frustration building on my girlfriend's face. It was probably, at most, three feet deep, so we stepped out.

No big deal; it was like an abandoned ship drill. We were Coast Guard and Navy veterans; how hard could it be? The new plan was to walk the canoe through the light rapids, then I'd run back the seven miles for the jeep. The thing I had forgotten was that my date had an issue with seeing blood—especially her own.

She stumbled on a rock and cut her leg. When she inspected the minor wound, it triggered a big problem; she immediately passed out. I rushed to grab her and let go of the canoe . . . which began making its way down the river with all our camping equipment, wallets, and yes, keys.

She wasn't a big girl, perhaps 110 pounds, but it was a dead weight. I finally got her to the bank where she came to her senses, looked at me, then her leg, and passed out again. Noting that she was in good shape—albeit unconscious—I swam down the canoe and managed to get it to the bank, find her keys, apply first aid to the leg and went after the jeep. I hitchhiked, which is not recommended in Appalachia, but we were in an extreme situation.

Needless to say, after a few stitches, that sealed the deal for us moving on to meet other people.

Training Center Yorktown is arguably the prettiest base in the Coast Guard. It's like a petting zoo with deer and other critters roaming without fear through manicured fields. Especially in the fall, the deer come into the parking lots. I'm told that when cars drive over and crush acorns, it makes it that much easier for them to get a tasty treat. In the spring, some will actually give birth on the parade field right outside the classroom windows. Teaching is hard enough with distracting students, but sometimes you just have to take a pause.

Getting to the base is fascinating if you like military history. Tucked away south of Yorktown, you are required to drive through the very battlefields where George Washington beat Cornwallis and where we battled to become an independent nation. We rode bikes through the cobbled stone roads at lunch and thought about what it must have been like 230 years earlier.

I reported to Hamilton Hall, named after Alexander Hamilton, George Washington's aide, and the first secretary of the treasury. He is also considered the founding father of the US Coast Guard.

After a long siege of the British, a final assault was needed on the entrenched British positions at redoubts #9 and #10. To take them with minimal loss of life required changing tactics, using stealth and a whole lot of courage. The rifles of the day were temperamental. In a large force of hundreds of men, chances were good that a rifle would accidentally go off, alerting the British to an assault. In a bold move, Hamilton volunteered for the mission and had his unit unload their weapons and make the assault with bayonets only, under the cover of darkness. They prevailed and paved the way for victory.

Redoubt 9 and 10 at Yorktown Battlefield

In a high-stress environment like combat, having a well-communicated commander's intent (take redoubt #10) and simplifying what the soldiers had to do is the key to victory. Instead of firing, then stopping and putting in wadding, then powder and then a ball, and then tapping it down, they used that pent-up energy to make the charge and know a bayonet couldn't misfire. There is much to be learned from our ancestors.

After the mission with the Navy, the School Chief Jim Loew knew I might get a little bored teaching in the repetitive Boarding Officer Course, and the school was looking for someone to eventually take over the role of chief of the International Maritime Law Enforcement Training Team. We just called it IMLETT for short.

The team was led by LT Rich Stoud and consisted of seven to eight folks: a couple of officers and chiefs, and a few petty officers. These were the primary deployers, and then we augmented other instructors from the rest of the school staff. This program was relatively new, and the mission was rapidly expanding.

A second Coast Guard team conducting international work was the Drug Interdiction Assist Team, or DIAT, headed up by my classmate Steve Baynes. Steve was a charismatic guy, graduated at the top of the class and was a Golden Gloves boxer before attending the Academy. During his tour leading DIAT, Steve was recognized for his cutting-edge leadership.

Because the Coast Guard was careful to protect its humanitarian image, Steve's team changed its name to IMLET with one T. The difference was Steve's team were all operators, and my team was all trainers.

The book *Not Your Father's Coast Guard; Untold Stories of Coast Guard Special Forces* by Matt Mitchell describes their efforts to destroy drug labs as part of the DEA's Operation Snow Cap. They were primarily focused on disrupting coca production, specifically in Bolivia, but had missions in Peru and Ecuador. I even get a shout out in the book for some Colombia activities, but that is a future chapter.

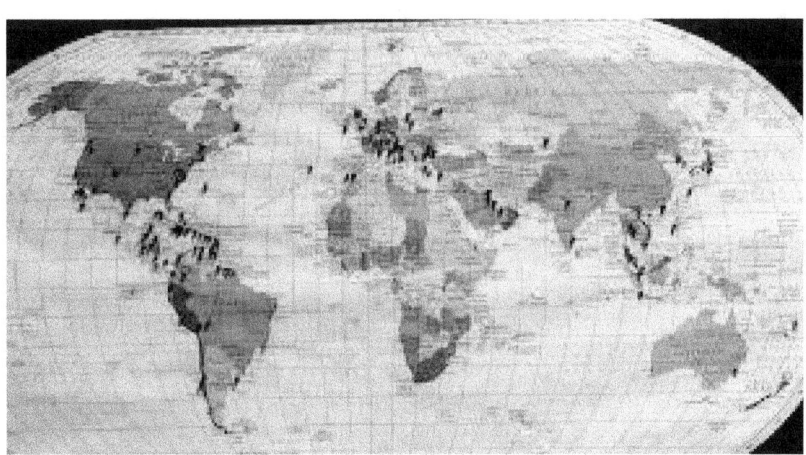

Marking Travel to seventy-six countries with the Coast Guard

The Coast Guard was well suited for Operation Snow Cap in the early 1990s. The highly televised Desert Storm demonstrated the advanced technology of our Navy. Most foreign navies found the US Coast Guard to be more closely aligned to their capabilities and defensive missions. Plus, we were still known as search and rescue experts and a humanitarian service, despite all the drug capturing efforts. Frankly, we were less threatening than our defense counterparts.

The international Coast Guard program, that included the IMLETT, was primarily funded by the Department of State or Department of Defense. Most of our funding and missions were focused on counternarcotics, but we also covered piracy and fisheries. As the team chief, I usually teamed up with an officer from Washington's International Affairs or a military attaché in the foreign country to conduct a pre-training assessment. This included establishing contact with the local nation's military or police to find out what they were doing, how best to tailor training, and to line up classrooms, boats, gyms and perhaps narcotics for testing.

I reached back to Yorktown and ensured we had course books translated and interpreters hired. Most of the students were navy or coast guard, but we also frequently had customs and national police. It was quite the adventure over the next three years. I filled and replaced my passport a couple of times because I had so many stamps and visas.

To be clear, while we were the instructors, these trips were an exchange of ideas. We probably learned more from the students than they did from us. We captured that information in trip reports and incorporated ideas and concepts into our domestic courses. Almost without exception, our in-country hosts were very gracious. The students were fun, and except for language barriers, they were very much like us.

I could probably write a book just on these trips alone, but I'll stick to only a few standout trips. As much as I was eager to go out, I was always excited to return to the United States.

During the latter part of this tour, I met my future wife Nye and started dating seriously. In my late twenties, many of my high school and college friends were married and had kids. Going to the bar with the guys was getting old. I'm not going to sugar coat it here; when we moved in together, the first visit by her parents was awkward. I scored points with the future in-laws when I brought home souvenirs from Thailand. They had met when Jerry was deployed to Thailand with the US Air Force during the Vietnam War. Nye had been born in Chang Mai.

For those who make it their life's mission to complain about how awful our country is, I wish I could invite them on these international trips. Some people lived in absolute squalor and amongst warring factions yet were very warm and patriotic. We have much to learn from our global neighbors, but that said, we still have the best country!

Lessons Learned:

Military personnel should know history. Alexander Hamilton used unconventional tactics, the cover of darkness and stealth to seal the deal for victory in the Revolutionary War.

In a stressful environment like combat, communicate clear commander's intent and simplify tasks.

One of the core missions for Special Forces is to train foreign internal defense forces.

The US and the Coast Guard were generally well-thought of around the world, especially in the early 1990s.

The *quality* of overseas accommodations and air travel vary greatly. At one point, the team was on the tarmac when the plane caught fire. The guys got to test the inflated ramp.

If you want to make a big impact, suck it up and go global. Don't forget to enroll in frequent flyer programs.

If you think your girlfriend will pass out by the sight of blood, buy a Kevlar or aluminum canoe.

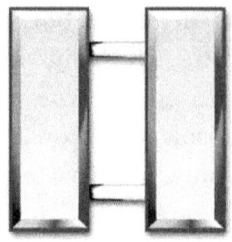

Chapter 9

Bolivia, Poland & Atlantic Area Commander

IF WE'RE GOING TO Bolivia, I need to start with a shout out to our former counterparts in DIAT, who later changed their name to the International Maritime Law Enforcement Team or IMLET with one T. My team was IMLETT with two Ts, adding the distinction of being a "training" team.

Michael Mitchell's book *Not Your Father's Coast Guard* focuses on this unit's activity primarily in Bolivia and their role in the destruction of cocaine labs. It's a great book and will give you a sense of the unit's somewhat-clandestine missions.

Joe Rodriguez at DIAT

One of my DIAT friends—and later battle roster port security shipmate—was Joe Rodriguez who epitomized this team. Below is a picture of Joe from those days of high adventure. It was taken in Bolivia where the two teams met up.

Before I get into the specifics of the trip, it's prudent to give you the background of what brought us together in a landlocked country in the middle of South America.

Coca leaves, grown in Bolivia, were turned into coca paste. There was a

84

short time window to fly it north to more extensive labs in Colombia that converted the paste to cocaine. From there, it was packaged for further movement north to the United States and Europe.

From an administration standpoint, we spent a great deal of effort interdicting vessels and aircraft flying with small loads of cocaine. It wasn't really making the impact we needed. I had personal experience on what it took to interdict a few hundred pounds when hundreds of tons were moving and getting into our cities. If we in the Coast Guard were stopping hundreds of pounds in a bust, think about the undercover police officers risking their lives for a few ounces here or there.

If you're not winning the war, you change the strategy. The US State Department, in conjunction with the Department of Justice, devised a plan to counter the smuggling process closer to the source. The DEA's Operation Snow Cap was born. Essentially, the plan was to create a Bolivian force with surface and air components to eradicate these labs. The US organized, trained, equipped, and operated alongside the force, now known as the UMOPAR.

UMOPAR translates to the Mobile Police Unit for Rural Areas. This organization was created in 1987 as a subsidiary of the Special Anti-Narcotics Force of the Bolivian National Police. This Bolivian counternarcotics and counterinsurgency force was founded, funded, advised, equipped and trained by the United States government as part of its War on Drugs.

Although UMOPAR was technically headed by a branch of the Bolivian Interior Ministry, they were, in practice, controlled by DEA and US military officials based at the US Embassy in La Paz who planned their operations, provided intelligence and led the drug raids, using UMOPAR mainly as a strike force for US operations. UMOPAR forces received extensive training from DEA and US military personnel.

In 1987, under a US State Department contract, several private military contractor pilots, many of whom had flown for the CIA's Air America in Laos and Cambodia, transported DEA agents and UMOPAR troops throughout the Upper Huallaga Valley in Peru.

In 1988, US Ambassador Rowell decided that UMOPAR troops needed their own airmobile task force to increase effectiveness. The US Department of Defense loaned UMOPAR twelve helicopters. Rowell assigned his US Army-Navy attaché, Lieutenant Colonel Edward Hayes, to command the UMOPAR troops in the unit called the *Diablos Rojos* (Red Devils).

In addition to troop transport, there was also an air operation known as Air Bridge Denial where the United States' allies shot down coca paste planes. After an accidental shootdown of an American missionary and his daughter in 2001, that program was suspended.

Beyond the coca growing, perhaps you're asking why the Coast Guard deployed to landlocked Bolivia? Despite having no coast, Bolivia does have a navy, nicknamed the Blue Devils, comprised of River Queen-type mother ships and a small flotilla of outboard boats called piranhas. The small boats were appropriately named as they operated in the headwaters of the Amazon.

Trinidad River Queen

If this law enforcement team didn't have precise coordinates of a lab, a common tactic was to load the small boats with UMOPAR personnel and drift silently down the river. If they heard the hum of a generator indicating a possible lab, the force assaulted from the water, land and air.

The IMLET and their DEA counterparts went on these missions to ensure intelligence was collected, arrests were made and the facilities were destroyed. When the US-aligned law enforcement team mounted a raid on a town that was a drug organization stronghold, elements of the Bolivian Navy (Blue Devils) were shooting at the Bolivian Air Force (Red Devils). I will not get into too much detail other than to say there were Coast Guardsmen involved in this assault force, and they had to restrain a Bolivian naval officer who would have blown the whole operation. Numerous boats and aircraft were seized.

It was the IMLETT's (with two T's) turn to conduct training with the Blue Devils and UMOPAR. We flew into the capital city of La Paz in the Andes Mountains. La Paz sits at 11,942 feet above sea level and is the highest capital in the world. Before flight attendants open the aircraft's door to deplane, they must lock the oxygen masks or else they would immediately deploy. The only time I had been higher was on my Boy Scout trip to New Mexico when we climbed snow-capped Mount Baldy.

La Paz sits at 11,942 feet above sea level.

Because we flew in from sea level, we were immediately winded, and headaches were not unusual. The Bolivians didn't need a fire department in La Paz. Number one, they were above the tree line, and wood structures were rare. Number two, there was too little oxygen for fire to be a threat. That was what the hotel gentleman told me as he poured me coca tea designed to help my splitting headache. The tea made from coca leaves would have relieved my headache but would have another unwanted side effect: a positive test for cocaine. That meant looking for another career, so I declined the offer and wolfed down several aspirins.

After meetings with embassy personnel, we headed back to the airport and prepared for the flight to the training site at a much lower elevation. At the counter, a young lady told us we couldn't fly to the Bolivian town of Trinidad due to the rain. First of all, Trinidad is in a rain forest. Second, we had flown this far, expending a lot of money.

I briefed our situation to the embassy and International Affairs, and they told us to do what we had to do to get to Trinidad. I then offered the nice lady at the counter an airport tax of $50 (USD), and magically, the weather improved.

As we flew northeast, the arid mountains gave way to savannah and eventually became a lush, green, rain forest. We needed to get to Trinidad, which was near a river and the Bolivian naval base. Trinidad was the forward operating base for Operation Snow Cap and contained a hotel that housed both the IMLET and DEA. The town of Trinidad was an uneasy place to live. It felt like a Wild West town where a gunfight could break out at any time. After all, Bolivia was where Butch Cassidy and the Sundance Kid met their fate!

There were effectively two sides: the Americans and the UMOPAR, or the Narcos. Both sides knew the other and where they resided.

The hotel housing the Americans was not five stars. But it did serve its purpose as a fortress with bodyguards. Many critters creeped about in the Amazon, and inconveniently, there was a half-inch gap below our doors. When we shut off the light at night, it only took ten to fifteen seconds before numerous glowing eyes appeared. I learned to leave the light on. It was also a good idea to shake out our boots in the morning.

If you've seen the movie *Romancing the Stone*, remember the scene where they were headed into the jungle? It looked like that outside the small town. When we were driving in on the dirt road from the airstrip, the driver anxiously pointed out an extremely attractive lady in an outdoor shower. It must have been laundry day.

These people were dirt-poor, subsistence farmers. Children, chickens and dogs darted about. Despite not having material goods, they were doing what kids do best: playing and laughing.

Our classroom had a similar ambiance. Our Bolivian host had power, but adaptors were needed. We taught through interpreters, and a lot of what we did was hands-on, like arrest and search tactics. At one point, I had to go to the store for classroom supplies with one of the DEA agents. He passed me a roll of money to hold while he negotiated a purchase in Spanish. It was $5,000 (USD). I don't think I've ever again held that much cash in my hand. That kind of money was used to encourage sources to help locate labs or specific individuals involved with cocaine production. This was a brave new world, and being relatively young and less than street smart, I didn't fathom how easy a mark I'd be for mugging.

We had a cover story and carried tourist passports, in addition to official ones, in case we were stopped at non-government checkpoints. The official passports were stamped by some bureaucrat in Washington, clearly stating that our purpose was counternarcotics. Not something you wanted to advertise in South America in the early 1990s.

Our training on the river each day hinged on the weather. If it rained for more than twenty minutes, we camped on the river, which made our hotel rooms look like the Ritz Carlton. When the *Jungle Queen* and UMOPAR small boats headed out on a mission, they weren't even issued food or provisions. They hunted, fished and essentially lived off the land. One such task force got lost for several weeks, and all communications ceased. When they finally emerged, they had destroyed a record-breaking number of labs.

When the South American part of our mission was over, Petty Officer Brian Smith and I flew from Bolivia through Miami to New York and then to Warsaw, Poland.

It was great to be back, albeit briefly, in the States. On the flight to New York, Brian and I were ready for yet another leg of the journey when the tall, distinguished Coast Guard vice admiral boarded the plane in uniform. Vice Admiral Welling was the current Atlantic Area commander. Mike and I were not in uniform, and I just nodded when he passed. He had no idea who we were.

It was the first of the month, and Mike was officially authorized to put on E6 or first class petty officer. In anticipation, I had purchased "crows" before leaving Yorktown and had every intention of presenting them on the plane. But I got a better idea. I passed a note to the flight attendant to give to the vice admiral. It briefly explained who I was, and that Mike was promoting. I asked if he could possibly present the crows to him as we disembarked in New York. He agreed, and in New York, he held up in the breezeway and surprised Brian.

Afterward, he asked what we were doing. He had no visibility of our mission in Bolivia as it had been entirely run from Washington. We told him we were heading to Poland for the next mission. Both countries were in his area of responsibility, and he didn't know about Poland either. The Duke, as they called him because he looked and acted like John Wayne, thanked us for our service, congratulated Brian, and bid us adieu. We moved to the next gate.

When we got to the gate for Warsaw, a page rang over the intercom. "Lieutenant Ogle, please report to the gate for a message."

I was used to going a little incognito, so I was surprised when I heard the rank. On the way to the desk, I honestly thought I screwed up with providing too much information to the Duke. But that wasn't it. It was an urgent call from the other team members supposed to meet us for the flight to Poland. Their connection had been grounded due to weather, and Brian and I had to fly on ahead.

Our arrival in Warsaw was a big deal. Poland was in the process of converting their missile boats into coast guard cutters to perform rescues and patrol their fishing grounds in the Baltic Sea. We were met at the airport by a Polish admiral who had his helicopter standing by for further transport. This was just a couple of years after the Berlin Wall had fallen. Poland had suffered under the Nazis and Soviets and was thrilled to have

Americans onboard, even a lowly lieutenant and freshly promoted first class petty officer.

We explained the flight delays and connected with the embassy, our interpreter and students. Our counterparts caught a later flight, and we departed Warsaw for the coastal city of Gdansk on the Baltic Sea.

As was the tradition with every course, regardless of the country, the locals always wanted to throw a post-graduation dinner party at a fancy restaurant for us. It, almost always, involved imbibing, except for some Muslim countries. But in Poland, we experienced something cool—a big bonfire in the woods.

Without a doubt, these guys loved their vodka. I admit, I'm very bad with languages, and I only remember one word repeated many times, *nostrovia*. That translates roughly to "let's get drunk." The vodka tasted like Everclear, burning all the way down. The key was to sip slowly and, when no one was looking, pour some out.

They were wonderful hosts, but when I got back to our barracks, which I believe was once a Holocaust site, I only remember dragging my forehead along the wall to find my room. This was the most intoxicated I had ever been. But it was my duty!

Lessons Learned:

Always carry a tourist passport in case they stamp DEA in your official one. Have a cover story.

Sometimes a bribe or "airport tax" is necessary to complete your mission. Do this carefully with top cover.

Don't assume all senior officials in your organization are privy to all missions.

Eastern European countries recently freed from Soviet control love America.

When socially required to drink straight vodka, secretly spilling some is preferred to alcohol poisoning.

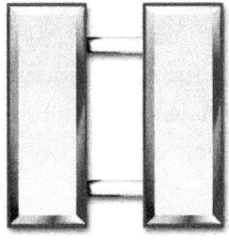

Chapter 10

El Salvador is at War, and We're Going In

IN FEBRUARY 1991, WE got word from Headquarters that a boarding officer course was to be delivered in La Union, El Salvador, on the southeastern coast. The fact that they were at war was not a showstopper.

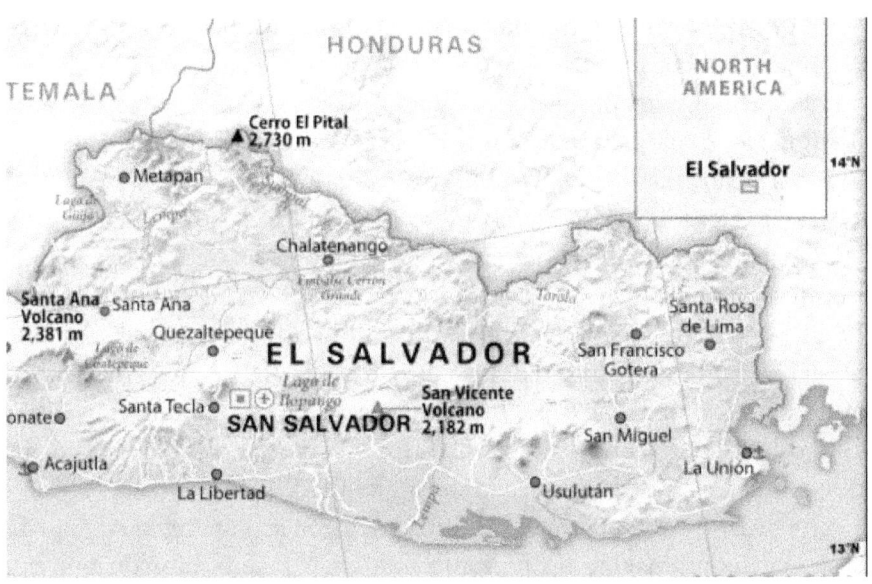

Central America had been in the news in the mid-1980s for the Iran-Contra Affair. This was a political scandal in the United States occurring in the Reagan administration's second term where senior administration officials secretly facilitated the sale of arms to Iran's Khomeini government, which was subject to an arms embargo. The administration hoped the proceeds of the sale could fund the US backed, right-wing

Contras in Nicaragua. This was part of the strategy to stifle the spread of communism in the region. There was no doubt that El Salvador was a hot spot for violence. You could skip a rock in La Union and hit Nicaragua. Its other neighbors, Honduras and Guatemala, were in a fragile state with growing left-wing insurgencies.

The US had just prevailed over communism in the Cold War and was eager not to see a resurgence, especially in the western hemisphere. The Salvadoran Civil War was fought from October 1979 to January 1992 between the military-led Junta government of El Salvador and the Farabundo Marti National Liberation Front, a coalition of left-wing groups.

The fully fledged civil war was in its twelfth and final year, which included the deliberate terrorizing and targeting of civilians. This was akin to the Vietnam conflict in that the United States found itself in a jungle with the intention of stopping the spread of Marxist ideology. This put the United States, again, in alignment with right-wing government whose tactics often violated human rights. It was messy on both sides.

Like many conflicts, this put us as trainers in a precarious position. An example would be our teaching on non-lethal, law enforcement, pressure points to help control non-compliant protesters. Those same pressure point tactics could be used for different outcomes, and therefore we had to modify our training.

To be clear, our role was to fly to the eastern naval port city of La Union and train the El Salvadoran Navy in counter-narcotics smuggling. Luckily, our team secured a weeklong terrorist awareness course in Fort Bragg put on by US Army Special Forces in advance of this mission. Designed for those being stationed abroad, we even had an Army general heading to Egypt in the class. I think our classmates found having Coast Guard in the class unusual. I've had dozens of schools throughout my career, but this was the best one I ever attended.

As part of the training, counter-intelligence guys tracked us off base—including paging us at Applebee's—put fake bombs on our car, and trained us on how to operate a variety of weapons should a terrorist go down, and you needed to use their weapon. We even had car driving, which included "J" turns and crashing the car through barricades.

In El Salvador, there was an embassy requirement that all military personnel in-country carry concealed weapons for personal protection, so we made sure to get our certification current. I bought both a shoulder holster and a European man bag worn around your waist with a tear-away pouch for our standard 9mm pistol. This was a civilian clothes' trip for sure, as the threat was high.

On the flight down, I recounted the Shenandoah River disaster with my teammates and how it was difficult to meet and keep a girlfriend with our deployment schedule. I was starting to consider a long-term relationship, perhaps even have kids. One of my traveling companions, a chief, pulled me aside and said, "You know what your problem is? You look like Joe Shit, the ragman. Where did you get these clothes? A thrift shop"

I had been working out to put on some badly needed muscle, which resulted in me outgrowing some of my few civilian clothes. It's not like I had a wardrobe at the Academy, and it was hard to keep up with the latest fashion trends. The rest of the trip, they called me Joe or J.S.T.R.M. For what it's worth, in a *subtle* chief way, I knew I had a mission when I got home. If I wanted to be a contender with the ladies, I needed to hit JC Penney's.

Excitement started as we landed at the airport in the capital city of San Salvador. We made it to the customs checkpoint, where they x-rayed our normal, large, black, plastic, pelican cases. As these boxes went through the machine, almost immediately, the operator got nervous and made a radio call. Within seconds, we were surrounded by soldiers pointing rifles at us and yelling in Spanish. This was when I wished I had paid more attention to that one year of Spanish in high school. The x-ray machines had picked up the silhouette of our aluminum non-guns, handcuffs, and batons that we used in training.

We had not yet met up with our interpreter or embassy representatives, and the next couple of minutes were tense. I repeated with my hands up, "Falsas Armas. Estados Unitas policia."

Upon inspection, they nodded and lowered their guns. Finally, the embassy and Salvadorian military arrived and let them know we were expected. After a quick change of underwear, we moved to the curb where we were met by a hermetically sealed, black, bulletproof, embassy suburban.

The soldiers, now allies, were clearing the way for us. We were given a very brief welcome in the truck that included some important safety tips if we were ambushed before arriving at the embassy. Unlike the briefing by the flight attendants going through a seatbelt's locking mechanism, I actually paid attention to this one. At the embassy, we met our handlers and received our 9mm pistols. They were always to be concealed and carried on our person. This was not the normal, warm, friendly reception that you'd expect at most embassies. There was no time for pleasantries. They were all business and earning their paychecks.

At the hotel, per our training, we surveyed all escape routes and always ate in the position to surveil the restaurant and street. The next morning, our black suburban arrived, and we were transported to a large soccer field in the middle of the city. Two Huey gunships with door gunners swooped in and landed. It was like a scene out of a Vietnam movie. We embarked the helicopters and were directed to sit on our helmets as the helicopters would occasionally take ground fire. I was thinking, *Holy crap! I might need a bigger helmet.*

Knowing the helicopters fly fast and low, the rebels put IEDs in the

The team travelled to La Union via Huey Gunships (photo courtesy of DVIDs)

treetops on known routes. The safest way to get to La Union at that time was to fly due south to the coast, then fly east over the Pacific Ocean. The last leg was over rebel-held territory, so the door gunner stood at the ready.

When we landed at the La Union Naval Base, we were escorted to a small American compound within the base. It looked like a Vietnam War

compound. One of the first things they showed us was how to set off claymores mines if the rebels made it to the fence line. Long guns were stashed under every bunk. They said, "If you see some guys that look American, don't worry; they are just electricians."

Okay. Right.

During one of our classes, the rebels did attack and blew the power to the base. Our hosts explained it was a regular occurrence, and that's why their American generator was needed. I think if our headquarters had known what it was like on the ground, they might not have sent us.

We did have a good time with the other Americans. Most were

retired military of some sort, and we got the typical Coast Guard jokes. I was well prepared after my 2nd Fleet deployment with a snappy counter.

One night after class, some of the students and other Americans sharing the compound invited us to go into town for candy bars. When we agreed, everyone grabbed a machine gun, and we were off in a bounding overwatch formation. That was a great candy bar.

Apprehending "bad guys" during an exercise

As for some retrospective thoughts on El Salvador, the US Government bought gas for their Navy boats to conduct counter-narcotics patrols, but word on the street was that those same patrol boat captains would sell the gas to local fishing boats and pocket the money. But reporting corruption like this can get you in hot water.

During a different international country visit, some of the students felt confident in sharing a concern with me about details of corruption in their ranks. I included it in my trip report. Despite the report being marked NOFORN (not releasable to foreign nationals), the embassy representative shared and then retired as an expatriate in that country. This is why I don't go back to that specific country.

Based on our experience with this trip, one can see how gangs like MS-13 came into existence. A very violent element was prevalent in the country. While I believe this was my most exciting international trip, I want to reiterate we were teaching the El Salvador Navy minimum force to compel compliance. That included the use of non-lethal pressure points.

I was always worried that those techniques might be used differently. El Salvador never became communist, but you must be wondering if our efforts were on the right side of history.

Lessons Learned:

If multiple guns are pointed at your head, raise your hands. Learn basic customs and phrases in advance.

If you are flying in a combat zone and expect ground fire, sit on your helmet.

Learn how to safely use a claymore mine.

While shoulder holsters look cool, go with the waist belt.

Not everyone overseas is what they say they are.

Study geography and terrain in advance in case your helicopter is shot down.

If you want to be a contender, wardrobe matters.

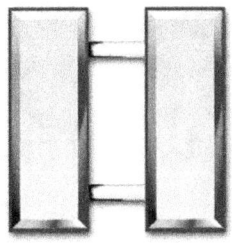

Chapter 11

What Country Are We In?

AFTER THE EL SALVADOR mission and a successful wardrobe shopping spree, I was ready to deploy again. This time to South America. Although after that Central America trip, I was starting to think Headquarters International Affairs was deliberately trying to kill us.

Peru was one of the critical source countries for coca production and therefore made it a prime training target for the IMLETT. IMLET (with one T)—formerly known as DIAT—also had a presence there.

Our mission was to fly the team to Lima and conduct a boarding officer course with the Peruvian Navy, so we decided to fly out early and have all day Sunday to recuperate from the long flight. I was not dating anyone at the time, and incredibly, I met a young lady that Saturday night, and we spent the next day together. It must have been my new clothes.

The embassy warned us that a Peruvian admiral would be kicking off the course Monday, so we needed to be rested and excited to get going. I woke up early Monday morning with the normal course kick-off jitters, concerned about details like power converters, interpreters, and base access. In the States, Murphy's Law is real. It is out of control overseas. In the hotel room, I turned on CNN to breaking news with tanks and troops in the street, tear gas, and a guy getting pistol-whipped by soldiers. I didn't think much of it and jumped in the shower. When I got out, it was still on, so I looked a little closer, and crap, it was a coup in Peru.

The images of unrest were in Lima, just outside the hotel. I contacted the embassy security official to see what our status was. The Peruvian admiral who had planned to kick-off the course was a "bit busy." He did pass on that it wasn't as dangerous as CNN made it look. The soldiers

Coup d'etat in Peru

didn't have bullets in their weapons, and the guy getting pistol-whipped was being shown in a variety of angles to create the illusion that it was more significant. There was some tear gas but only a few canisters.

Meanwhile, back at Yorktown, our commanding officer Captain Paul Pluta was tuned into the same CNN channel, so I was told that he was exploring the possibility of arranging a C-130 to get us out. Considering what we had heard from the embassy, from the Peruvian admiral, and the fact that we had met up with some pretty ladies, we recommended that we continue with the course and complete our mission. The violence on the streets of Lima was more or less over that first day, with the pinnacle being tanks and armored vehicles arriving at the building housing their congress and the use of tear gas to disband the governing body.

Despite trouble in the streets, the week-long course went well and leaving was more difficult than I expected. But as a young LT, I didn't have a good idea of what was actually happening on the ground. From that point on, I took advantage of country briefs, so we knew what we were getting into. Years later, with the advantage of hindsight and Wikipedia, I learned this:

The 1992 coup was a constitutional crisis after President Alberto Fujimori dissolved the Congress of Peru as well as the judiciary and assumed full legislative and judicial powers. The country was in a downward spiral under his predecessor who still had allies in the other government branches. They were blocking Fujimori's attempts to right the ship with sweeping reforms. Prior to his inauguration, Fujimori traveled to Japan and the United States to meet with top-level officials and request aid for Peru. While he was in the US, he was told that Peru must adopt a "relatively orthodox economic strategy" and stabilize hyperinflation before being permitted to re-enter the international financial community. Bottomline: without these policy changes there would be no international aid forthcoming. The Peruvian Congress, however, resisted Fujimori 's efforts to adopt policies advocated by the International Monetary Fund and World Bank.

We had arrived with our team on Saturday night, April 4. On Sunday night, April 5, 1992, President Fujimori appeared on television and announced that he was "temporarily dissolving" the Congress of the Republic and "reorganizing" the Judicial Branch of the government. He then ordered the Peruvian Army to drive a tank to Congress's steps to shut it down. When a group of senators attempted to hold a session, tear gas was deployed against them. That same night, the military was sent to detain prominent members of the political opposition.

There was little initial domestic resistance and in fact an opinion poll carried out shortly after the action had an 85% approval rating. The economic and political situation was so poor at the time that many Peruvians thought it could get only better, and Fujimori's bold and risky economic reforms appeared to be working.

Fujimori claimed that the coup was necessary to break with the deeply entrenched interests that were hindering him from rescuing Peru from the chaotic state.

International reactions were different: International financial organizations delayed planned or projected loans. The United States, Germany, and Spain suspended all aid to Peru other than humanitarian assistance. Venezuela broke off diplomatic relations, and Argentina withdrew its ambassador. Chile joined Argentina in requesting that Peru be suspended from the Organization of American States. The coup appeared to threaten the economic recovery strategy.

Even before the coup, relations with the United States had been strained because of Fujimori 's reluctance to sign an accord that would increase US and Peruvian military efforts in eradicating coca fields. This was ironically why our team was in country. Although Fujimori eventually signed the accord in May 1991, the disagreements did little to enhance bilateral relations to get desperately needed aid. The Peruvians saw drugs as a US problem and the least of their concerns, given the economic crisis, Shining Path guerrillas, and an outbreak of cholera.

Two weeks after the coup and one week after our team returned to Yorktown, the Bush administration changed its position and officially recognized Fujimori as Peru's legitimate leader. The Organization of American States and the US agreed that Fujimori's coup might have been extreme, but they did not want to see Peru return to the deteriorating state. In fact, the coup came not long after the US government and media had launched a media offensive against the Shining Path rural guerrilla movement. On March 12, 1992, Undersecretary of State for Latin American had told the US Congress: "The international community and respected human rights organizations must focus the spotlight of world attention on the threat which the Shining Path poses . . . Latin America has seen violence and terror, but none like this group . . . and make no mistake, if the Shining Path were to take power, we would see . . . genocide."

Given concerns, the long-term repercussions of the coup turned out to be modest.

One of my Peruvian students tried to explain "Indian Badness" prevalent in South America. The example he used was metal boxes that contain newspapers you find at a 7-Eleven. He said, in the States, a person puts a coin in and takes one paper. In South America, a coin goes in, and all papers are removed and sold on a street corner.

In a later trip to Ecuador, one of the students joked that they had a military strategy to declare war on the United States. They said the war would likely be over in less than a day, and then the United States would come in and completely rebuild their country better than it was.

The United States as a superpower often tries to apply our values, which have made us successful, to other countries and cultures. I would offer that generally doesn't seem to work. What works is supporting their autonomy, helping them help themselves, listening, sharing our lessons learned and being a beacon of freedom and opportunity.

Lessons Learned:

Unlike most movies, sometimes you want to stay in a country that is experiencing a coup.

Political and economic challenges vary widely around the globe. The Left or socialist groups *tend* to align with terrorist and narcotics organizations. The Right with capitalism, military, and police. Sometimes our democratic standards don't work well.

Sometimes the media can overblow what is happening on the street.

Get a current threat assessment from the CIA or Department of State before traveling overseas.

The President of Peru was of Japanese descent.

Chapter 12

Six Weeks in India, Singapore and Malaysia

PERU HAD BEEN A long trip but was at least in the same time zone as Virginia, due south and across the equator. If it's winter in the Northern Hemisphere, it's summer in the Southern Hemisphere. That was *really* important to know when upgrading my wardrobe and packing for Chile, Argentina and Uruguay, but not so much for Lima. Lima had been the same distance south of the equator as Puerto Rico had been to the north. It was tropical near the coast.

On this trip, New Delhi—the same latitude as northern Florida—was the first stop on our trip to circumnavigate the globe, always traveling east. Singapore is just over one degree of latitude, or roughly seventy miles, north of the equator, and Malaysia was farther north. Therefore, the whole trip was in the northern hemisphere, but all in the tropics.

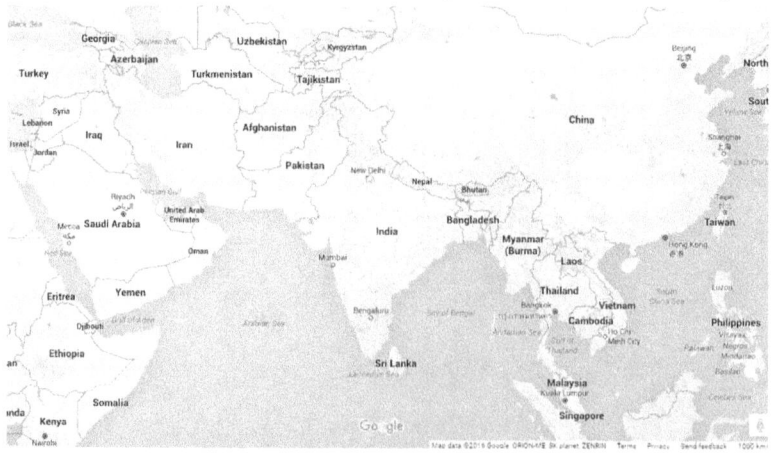

If you had asked me when I was a boy scout where I would like to travel, I might have said India, because it was on the other side of the world. It was exotic with Mount Everest just to the north in Nepal, and to the south, it was covered in jungles and home to monkeys, elephants and tigers. It was also home to one of the seven manmade wonders of the world, the Taj Mahal.

India was one of the few countries aligned with Iraq during the Gulf War just a couple of years earlier, so the fact that they were interested in entertaining training from the US Coast Guard was somewhat unexpected. It had been over thirty years since a US military team had trained in India, so this trip was a big deal. India is the second most populated country in the world and has the seventh largest economy.

This was certainly different from my normal trips involving smaller nations. India had a sizeable navy that included aircraft carriers, but they also had a formidable coast guard. They were very much interested in countering drug and gold smugglers and had a festering terrorist problem. The Tamil Tigers of Sri Lanka were some of the most sophisticated terrorists globally and even had their own maritime fleet that included submersibles. When the government forces finally got the upper hand on the Tamil Tigers, the engineers who built the submersibles moved to South America and started working for the drug cartels.

We were to conduct the training on on a naval base on the India's southeast coast just west of Sri Lanka. This mission had a lot of visibility up the chain and was the typical two-part trip: the assessment and follow-on training. Because of the visibility, they sent the big dog with me—Jim Loew—for the first part.

As you can imagine, the jet lag was quite significant after nineteen hours in the air and the nine-hour time difference. After a brief down day, we met with the embassy and Indian officials in the capital city of New Delhi. At this point, I was comfortable doing assessments, and Jim and I quickly constructed a plan for the follow-on training with the rest of the team arriving in a couple of weeks. The fact that India, like the United States, had been a former British colony meant English was prevalent. We not only taught in English, but our course materials didn't need to be translated. But language and the modern naval force was about the only similarity.

We had some time to explore, so we headed into the big city with a population of twenty-one million. New Delhi was teeming with people riding on all sorts of conveyances ranging from cows to stretch limousines. The most common was the three-wheeled rickshaws. Apparently, they didn't have effective environmental laws, and the city hung in a perpetual pollution cloud.

The Hindu religion had many different gods, thirty-three million to be precise. While not one of the big four, the one I most frequently saw was Ganesh, an elephant-headed Hindu god of the beginnings, who traditionally is worshipped before any major enterprise such as running a training course. Ganesh is the patron of intellectuals, bankers, scribes and authors, so I asked the class to put in a good word for me. About 80 percent of India's population is Hindu, 14 percent is Muslim, and the remaining small percentage is Christian and Sikhism.

Driving through the crowded streets of New Delhi was heart-stopping, especially getting Jim to the airport. On one occasion, our driver had to decide: swerve left and hit a man on a bicycle, or right and possibly hit a sacred cow running free in our general direction. We swerved left, fortunately missing the bike.

Driving in New Delhi is not for the faint of heart.

There was a caste system in India. Think of a pyramid with a cloud over the top. The cloud is where the Gods reside. Then moving down to the peak of the pyramid, you have the Bhramin. These are the priests and academics. The next layer down is the Kshatryia—the warriors and kings. The next level down is Vaishya—merchants and landowners. The next level down was Sudra—commoners, peasants and servants. Then there were the untouchables. They aren't even considered part of the pyramid. They are below. They are the outcast of the caste—street sweepers and latrine cleaners. Technically, Academy freshman were in that untouchable group. Marriages were mostly among the same caste and religion and often arranged by parents with or without the newlyweds' consent.

I'm all for freedom and equality among genders and races, and I think we have done wonders in the United States. That said, there are occasions when we need to be somewhat discriminatory when we are guests in a different nation. There are some in the US who say we should send a female instructor to demonstrate how western democracy has evolved, but in some places, that simply wouldn't work.

It would not have been fair to the female instructor or the class or beneficial to bringing the nations closer together had we forced a female trainer on the team. India didn't incorporate women into the military until 1992. It was 2008 when women were commissioned, which was only in legal and education positions. In 2020, they expanded into eight more corps. Just as equality took time to evolve in the United States, it is evolving overseas as well. In a country of a billion people, who's to say what works for them. I think India would be a prime country for our US protesters to visit, but I wouldn't want to do that to India. It would be eye-opening to them.

Abject poverty was prevalent as you moved away from the center of the city. Parents along the street offered their children to us. Disease was everywhere. Fortunately, I think we had received every shot known to man before our trip.

There was such a huge disparity between the haves and have nots, the top and bottom of the pyramid, even greater than found in America. My experience in the United States with those of Indian descent, for the most part, had been very positive. They were always at the top of the class, warm, friendly, and hard-working. I'm told by those in India that these American immigrants were part of a brain drain phenomenon. Some of India's best and brightest saw that the United States offered greater potential for their families. Those who could afford it attended a US college and never returned.

The last night before Jim was to fly out, we attended an embassy party serving traditional Indian food. It was quite hot but also tasty. They pretty much put curry on everything. Gunfire erupted nearby at one point, and everyone, including the military attaches, ducked for cover. After the brief excitement, Jim and I got milkshakes to cool our burning mouths. I am pretty sure now that if you get served a milkshake with a gritty texture, you shouldn't drink it, but we were doing our best not to offend the host.

That next morning it hit, and it hit with a vengeance. Not to be overly graphic, but my boss and I had it coming out both ends simultaneously. The worst thing was Jim had to board a twenty-hour flight home. I felt bad for him as I camped out in the hotel bathroom for the day.

This started my solo week in India where I waited to fly to Madras to link up with the rest of the team. After recovering from the milkshake episode, I received a call from the embassy informing me that several Indian Coast Guard admirals were interested in having a follow-up meeting with just me. The embassy staff coached me a bit, and then I was taken to the Indian Coast Guard headquarters.

I was only a lieutenant, but they absolutely pumped me for intelligence. They were also extremely interested in the relationship between the US Navy and the US Coast Guard. It was apparently not that different than what we experience in the United States. After leaving the meeting, I reported to the embassy and briefed my handlers. My official duties done, I mentioned I wanted to see the Taj Mahal in Agra, about a four-hour train ride south.

They approved the solo trip, and I headed out on my own. Fortunately, they speak English pretty much everywhere. The toughest thing was money conversion. The Indian rupee equaled .013 dollars. As an example, a $50 US round-trip train ticket cost me 3739.11 rupees.

The four hour train ride was worth seeing the Taj Mahal in Agra.

The Taj Mahal is an ivory-white marble mausoleum that sits on the southern bank of the river Yamuna. It draws seven to eight million visitors a year. Its construction was commissioned in 1632 by Emperor Shah Jahan as a tomb for his favorite wife, Mumtaz Mahal. I always wondered what he gave his least favorite wife.

The train ride's actual cost was $6 US, and I might have been overcharged. The smell of the train was quite memorable. The cars were packed with people with barely room to stand. It's very important to secure your valuables and passport deep in your clothing or carry bag in these very close quarters.

But the packed train car wasn't the main source of the smell. In the morning, literally hundreds of people come out of their villages along the tracks for their morning constitution. After four hours, I was excited to pull into Agra and be off the train. There it was. The Taj Mahal.

It looks great in pictures but seeing it in person is quite a different thing altogether. The white marble changes color depending on the time of day, and at sunset, it's striking. I was able to tour this magnificent structure and purchase a replica model and other souvenirs for my family.

Upon my return to New Delhi, I finalized arrangements, flew to Madras, and met up with the rest of the team. Madras was substantially hotter than New Delhi, and we conducted most of the classes in a helicopter hangar aboard a naval vessel.

The class discussions demonstrated the gap in our cultural thinking. Some of the students recounted with glee a recent seizure of gold smugglers off India's east coast. Like our drug cases, the smugglers, when hailed, refused to stop. The Indians said they opened fire with their machine guns and eventually gained compliance. Their gunfire had inadvertently killed three in the village on the other side of the smuggler, but that was okay as they were able to stop the vessel that was smuggling gold. If that had happened in the United States, we would have protesters lighting fires in every major city.

India had been an adventure and an endless opportunity for souvenirs. The issue was how much I could carry with the airline weight limit. This was my longest trip of the whole tour, so I had packed heavily, and this was just my first stop. I pushed the limit leaving India but had some unique gifts for the family.

The two weeks of training went well, and we received positive feedback. Ambassador Clark wrote the Commandant and said we had won the hearts and minds of the Indians. I'm pretty sure that was a bit of an exaggeration, but we certainly enjoyed the exchange. We received the Commandant's Letter of Commendation ribbon for our efforts.

But this trip wasn't finished. We boarded the plane and flew to Singapore for an overnight, where I looked up my Academy classmate Tan Tow Koon and his new beautiful wife. Singapore was the most advanced, cleanest, and squared away city I had ever been to. Singapore's laws were extremely strict, almost draconian.

For example, you could be arrested for chewing gum in public. Drugs carried an automatic death sentence. Don't screw around in Singapore! Tan decided to take me across the border to the Muslim nation of Malaysia for some rest and relaxation (R&R) and a nice dinner.

Then it was back on the plane for two weeks of training in Penang, Malaysia. We had taught a course in Penang earlier, and I even knew some of the students from the previous course. They wanted to take us out. To recap: when Singapore folks want to have fun, they cross the border into Malaysia. When the Malaysians want to have fun, they cross the border into Thailand.

These students were in the Malaysian Customs, so I called my contact in the embassy and said we were doing a border recon for the counter-drug course starting next week and gained permission to cross the border into Thailand. Thailand was an adventure and set me up for success with the future in-laws. It also gave me a connection to future Commandant Admiral Allen, who had years earlier commanded a Loran Station in Thailand.

After six weeks in Asia, I was ready to get home.

Upon completing a horrendous travel claim and trying to square away a few overdue bills, I was ready for the next adventure. That came soon enough based on the war in Serbia. But before I departed, I went to one of those websites and started sponsoring a child in need. He was from India.

Lessons Learned:

Do not drink gritty milkshakes, but if you do, find the nearest train track.

Sometimes to have fun, you need to cross an international border.

Malaysia has the biggest bugs I have ever seen.

Illegal drug possession, use or sales have significantly different penalties, depending on the country.

Gum trafficking in Singapore has a fine of up to $100,000 US dollars and two years in prison.

The United States, with its free society and unlimited possibilities, has attracted talent from all over the world. The "brain drain" helped us but hurt the contributing countries.

I am glad we don't have a caste system in the United States.

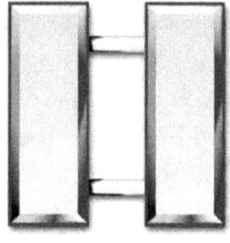

Chapter 13

Serbia, Romania and Bridge to Bulgaria

MY TRIP TO POLAND had given me a good idea of what the former Soviet Bloc's satellite countries were like; they embraced the west and cherished freedom that we often take for granted. The Poles and Balkan countries were consciously looking over their shoulder and feared independence could be short lived. The Poles had endured sixty years of occupation from the Nazis and then the Soviet Union. True freedom, marked by the fall of the Berlin Wall, was a new concept for the parents of those I was meeting. Violence and distrust ran even further back in the Balkans who were credited for the birth of the First World War.

When the Soviets left, Yugoslavia splintered into ethic countries: Bosnia and Herzegovina, Croatia, Macedonia, Montenegro, Serbia and Slovenia. When the new national borders were drawn there were winners and losers, which stoked friction in the region.

The Serbs, although initially militarily superior due to the weapons and resources provided by the former Yugoslav People's Army, eventually lost momentum as the Bosniaks and Croats allied themselves against the Serbs. Pakistan defied the UN ban on the supply of arms and airlifted missiles to the Bosnian Muslims. After two major massacres, NATO intervened in 1995 with Operation Deliberate Force targeting the Army of the Respublika Srpska, which proved key in ending the war. The most recent estimates suggest that around 100,000 people were killed, and 2.2 million were displaced. In addition, an estimated 12,000 to 50,000 women were raped, mainly carried out by Serb Forces, with most of the victims being Bosnian Muslims.

The Balkans were a powder keg. The United States had just come out of the Gulf War and was not eager to put more boots on the ground in a multi-side affair that had little United States strategic value. That said, we had also witnessed the atrocities perpetrated primarily by the Serbs. Limiting their weapons was one way to reduce violence. The UN had their weapons ban, as mentioned, but there was still weapons smuggling.

If you look on a map of Serbia, you'll see it's landlocked like Bolivia, so why send the Coast Guard? The mission was to serve as a military advisor to both Romania and Bulgaria. The Danube River flows east into the Black Sea and serves as the geographic border between those two countries. Following the river west, it borders Serbia. The concern was suspected arms shipments were being smuggled west on the Danube and fueling the conflict and allegedly contributing to ethnic cleansing.

In addition to suspected weapons smuggling, vessels transiting the river were taking fire on their pilot houses once they arrived at the Serbian border. My role was to see the situation first-hand and to file a report with recommendations for follow-on equipment, tactics and training to counter this river smuggling threat. The US Coast Guard possessed the most expertise in countering waterborne smuggling, and we now had an international training capacity. Both Departments of Defense and State had their fingerprints on this mission.

This was a solo mission for me, in the winter wearing civilian clothes. I was to intercept a representative from International Affairs in Bucharest and meet with the Romanian cabinet along with US Embassy personnel. As it was the most senior meeting I had ever attended, I was in the cheap seats.

The Romanian transportation secretary spoke surprisingly good

English, and I found Romanian to be similar to Spanish, so I could follow along most of the time. The meeting's conclusion was that the transportation secretary would drive my International Affairs colleague and me west along the river and get as close to the Serbian border as possible. Did I mention this was in the middle of the winter? My normal missions were in tropical Asian or South American jungles, so I was glad I had gone shopping for a nice ski jacket.

On the trip west, past Dracula's Transylvania, we hit blizzard conditions. It was so bad at one point, the car's fuel line froze, and we had to hike a bit to get it repaired.

When we got to western Romania, we found a hotel. I'd been to rough hotels before, but these post-communist facilities were a whole new level. There was a community bathroom for the rooms on the floor. No stalls, mind you. I filled a sink and took a picture. The water was brown, and their water heater was inoperable. Yet these people were very friendly, despite being neighbors to warring factions. I didn't realize before my departure that they may have been friendly to us, but that did not extend to their neighbor to the south.

My International Affairs colleague who had traveled the snow-covered countryside with us decided I could do the same mission by myself in Bulgaria. Back in Bucharest, after a debrief with the US Embassy staff, he boarded a plane back to the States. He must have known something I did not.

The embassy driver then drove me south to one of the few bridges that crossed over the Danube into Bulgaria. Many of the KGB enforcers historically came out of Bulgaria, so I didn't know what to expect. When I got to the Romanian side of the river, the embassy driver said, "This is as far as I take you. You will have to cross the bridge alone as we don't have the proper credentials. The US Embassy personnel on the other side will be waiting."

Honestly, that was unexpected, but I grabbed my suitcase, hopped a ride on what amounted to be a gypsy bus and got off on the other end. It was bitterly cold, and the river was iced over. When I hopped off the bus, everything was now in Cyrillic. J{o6pe ooutnu e EM2apu51,: Welcome to Bulgaria! (So much for English and Spanish working everywhere.)

To my displeasure, there was no US Embassy personnel waiting to meet me either. Remember, this predated cellular phones. I went through the Bulgaria customs checkpoint without a problem since I had the visa, then I found a payphone. Despite multiple attempts of speaking with an operator, I finally reached my contact who was running late due to the weather.

From there, I went to the embassy in Sophia and later met with Bulgarian military officials to tour their Soviet-era towers that lined the south side of the Danube River. Despite a feeling of distrust for westerners, they let me take pictures that helped fashion a strategy for detecting and boarding westbound vessels from the Black Sea heading toward Serbia.

Part of my recommendation was to fashion checkpoints on the river where the shipping traffic would normally have to stop, either at locks or moveable bridges. While waiting for the bridge to open or a lock to fill with water, these vessels are usually moored alongside the riverbank or required to wait in a queue.

Under this controlled situation, a boarding team can more safely inspect the vessel and use an explosive detection dog to greatly improve the chances of locating smuggled weapons. This tactic was nothing new for the United States. As a free society, we are ever weary of delaying commerce or using tactics seen as infringing upon personal freedoms. The US Coast Guard had been using logical stop points on the east coast intercoastal waterways to check primarily for drugs being smuggled northbound. Using natural stop points eliminates delays in commerce and unsafe boarding of a moving vessel in confined or restricted waters.

My trip report and recommendations made it up to our Joint Chiefs of Staff. It was a great mission, but I was eager to thaw out again in a Colombian Jungle.

Lessons Learned:

If you are going to the Balkans in winter, buy a heavy coat and thermal underwear.

Make sure you always double check your foreign contacts, handoff and contingency plans.

If you know a little Spanish, you can get along almost anywhere— except Bulgaria.

Balkan countries don't like each other.

The Danube River freezes over.

There are Soviet-era observation towers on the Bulgarian side of the river.

Don't assume just because you're traveling with the Romanian Transportation Minister, you won't have car trouble.

Add a dog to the boarding team.

If you don't want to disrupt commerce on a river or intercoastal waterway while making security checks, set up shop at locks or moveable bridges.

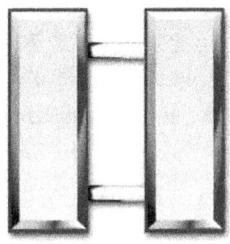

Chapter 14

Gomer Pyle, Colombia, and the IED

THIS WASN'T MY FIRST trip to Colombia. I had done an assessment in Bogota previously. I'd also been to Cartagena, which lies on the north coast.

Like La Paz, Bolivia, Bogota is a capital perched high in the Andes at a breathtaking 8,695 feet.

In the early 1990s, there were places in Colombia you simply could not go. The country was the epicenter of narcotic smuggling and narco-terrorism. The terrorist group known as the Revolutionary Armed Forces of Colombia or the FARC infested the countryside. Powerful drug cartel leaders such as the notorious Pablo Escobar in Medellin wielded power from major cities.

On the previous assessment trip, I went alone in civilian clothes and had permission to grow my hair out to look less military. I was annoyed that the visa issuing authority in Washington listed DEA for the purpose of the trip. That's why you always also carry a separate tourist passport.

While waiting in Miami's International Airport to head down on this mission, the seats were limited in the gate area. I had a cover story of what I was doing, but generally, my strategy was to put in earphones and avoid talking with anyone. I grabbed one of the last open seats in the waiting area next to an older guy, and just to be nice, I asked if he was on vacation.

In a southern drawl, he said, "No, I'm going to meet a cruise ship. I'm an entertainer."

"Like juggling or something?"

He laughed. "I'm a singer."

The more I listened to him, the more his voice sounded familiar. I caught a glance at his ticket, and sure enough, it read Jim Nabors. I was sitting next to Gomer Pyle from a show I watched as a kid. He was also in the *Andy Griffith Show* and was a resident of Mayberry. The stigma of hazing from the Academy suddenly came back.

I apologized, but I think he liked the fact that I didn't immediately recognize him. We talked for probably thirty minutes about his Hawaii ranch and the stressed-out Sergeant Carter. After boarding the plane, I gave him a head nod and moved to the cheap seats.

Senior people and even celebrities are often seen as unapproachable. Common folks don't want to engage as we suspect it would be perceived as annoying or even stalking. That was not the case with Jim Nabors. He enjoyed just being a regular guy, having a normal conversation with a stranger. I suspect that's true of many celebrities or senior leaders.

This Colombia trip was my last official overseas mission as part of this specific team. I had screened for command of a 110-foot patrol boat, and this time, it looked like I would get one.

While I was feeling fairly good about the future, this was not the time to get complacent. Back then, Colombia was considered the Bad Lands.

If you look at the map, I had already been to all the surrounding countries, including Peru, Ecuador, Venezuela, Guyana, Brazil, Uruguay, Argentina, Chile and Panama. The good news was Cartagena was considered a no-violence zone by the cartels and the FARC, as that's where their families vacationed.

There are beautiful women in the United States, especially my wife, but Colombia has perhaps the most beautiful I have ever seen. Even better, Colombian women, for some reason, like Americans.

Our team landed on a Saturday night in Cartagena. We met with the embassy folks who gave us the typical security brief. We also received the details of the course and our transportation arrangements. Quite a few embassy personnel showed up, and I think, due to tight security restrictions, they welcomed new faces.

We were off Sunday, and a couple of the young ladies from the staff asked if the four of us wanted to join them for a boat excursion the next day. Tim Cavanaugh, from the Desert Storm mission, and I were all in. The other two fellows said they were too exhausted from the flight and took a rain check. That was a huge mistake on their part.

We arrived at the local harbor the next morning, and sure enough, there was an old crusty America ex-pat guy with the boat and nine beautiful ladies, all pretty much in thong bikinis. I wasn't looking, but I think that old crusty American might have been wearing one too. It was something out of the movies, and we let our partners know they screwed up big time. We went out with the students from our course and our new lady friends almost every night. In Colombia, they partied *all* night long.

The course went well. These boarding teams had also received training from Israeli commandos. They were no joke. I believe the narcos underestimated the teams resolve to retake their country.

At the time, I really didn't think too much of it, but halfway through the course, I found myself walking down the street with a bag of marijuana (provided by the police) because we were doing drug testing that day. In retrospect, if I had been stopped by someone not given the memo, that would have been unbelievably bad.

By Friday, I was exhausted. Not so much from the course, but from being out at the clubs all night. It was a great final trip. I also came to the realization that my new girlfriend Nye, who I had just started dating back in the States, was a winner. A colleague had set us up on a blind date, and we hit it off.

She was Thai but had come over to the United States at a young age and was very much Americanized. Her stepfather had been in the Air Force deployed to the Vietnam War when he married Nye's mother.

My then-girlfriend Nye with her Colombian, emerald necklace

I decided to head out solo from the hotel to pick up a nice souvenir for Nye. As I walked down the street, a crowd formed around a corner of a building at the next intersection, complete with shouting and a police presence. The hair on the back of my neck started to stand up. Something told me this gringo needed to turn around.

Right after I turned around, there was an explosion in the crowd, and debris flew close to where I was. I jumped behind a parked car on the street. Uninjured, I quickly made my way back to the hotel, briefed the guys, and opted to remain in place until we departed the next day. Fortunately, they had a nice gift store in the hotel where I managed to get a nice emerald necklace for Nye.

After doing a little research (thank you, Wikipedia), the bombing was believed to be by the FARC, which is a Marxist revolutionary guerrilla force engaged in an armed struggle against the government.

It was time to move to Savannah and command the *Key Largo*.

Lessons Learned:

Simple acts of kindness can go a long way.

If there is an explosion, take cover. Remember the pressure wave goes out, and then it comes back.

If you are the only gringo walking down a Colombian street and you see a mob with police cars, turn around.

If you have a girlfriend, get her a nice emerald necklace found only in that country. Tell her you risked your life getting it.

If single and invited by pretty ladies in thong bikinis to go on a day sail, the answer is always, "When do we sail?"

The Colombian cartels have a cease-fire zone in Cartagena. They need to tell that to the FARC.

On long flights, make a list of life's goals...command, marriage, kids...write a book.

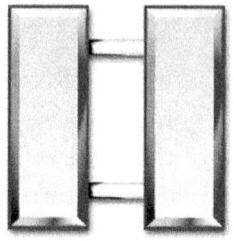

Chapter 15

First Command as an Officer

YOU NEED TO BE careful when speaking with the detailer. That's what we call the assignment officer in the headquarters who determines what job you get and where you're being sent. It's no doubt a very tough job where you make both fans and enemies, not only with the members but also their families. After retiring, I had former assignment officers as my students. They usually requested to be roommates, for self-preservation, I guess.

Assignment officers weigh your personal job request, which is submitted in the form of a *dream sheet*, with your experience, performance and the receiving command's desire. Service needs always trump the member's personal desires. Ask any military person you know, from a recruit to an admiral or general; they will tell you these conversations can be a delicate dance. You don't want to be needy or annoying, but you also want to make a case for where you want to go and what you'll be doing. It's not only how well you perform; those assignments determine your longevity in the service. If you want to continue to advance, you ask for the most challenging, high-profile jobs—like command.

I was thrilled to have been selected by the panel for command of a 110' patrol boat, but there were fifty of them scattered throughout the country, and screening doesn't guarantee you'll get a boat. The panel generally screens twice the number of people they need for positions available.

Back in those days, the correct response to the detailer is, "I'm honored I screened, and I'll go wherever the service needs me."

My dream sheet focused on the southeast United States, which would keep me near North Carolina where my father was battling terminal cancer.

Tybee Island was a barrier island off Savannah, Georgia, and was the closest boat. Fortunately for me, Jim Tunstall was my detailer, and by luck, he also came from the greatest state in the union—North Carolina. He was most gracious and assigned me to the Tybee-based cutter, the *Key Largo*. That was just a five-hour drive to Raleigh. That assignment paid off, and I was closer to home as my father's condition continued to deteriorate.

Key Largo was homeported in Savannah Georgia, but shown here in Boston.

When I took command of the *Key Largo*, I experienced something that was entirely new; this was my third ship, yet the first time we had any women aboard. Our XO was LTJG Laura Dickey, and she was extremely squared away. My only complaint was that she liked to play Michael Bolton songs. I had had my fill with that guy as Nye liked him too.

My predecessor had opted to run the patrol boat like a large cutter with standard workday routines. On *Vashon*, we didn't use that technique, but I gave it a shot for about a week. Patience is what they teach you in command school. You don't come in and start changing things until you get the lay of the land. If you have a strong desire to make changes in the first few weeks, move the furniture around in your office. My office on the patrol boat was smaller than most closets.

After a week of trying to run a small patrol boat as if it was an aircraft carrier, we suspended normal workday routines and loaded fishing poles. When not directly engaged in missions such as search and rescue or law enforcement, we relaxed. Like a fire station, we had to be ready to roll when the bell rang, day or night, so we rested when we could. On small cutters like the *Key Largo*, almost all missions required everyone to be up and working, hence the phrase "all hands on deck."

Laura was a great XO and had learned from some of the best. She quickly adapted to my style and supported the relaxed schedule.

When I was writing this book, Rear Admiral Dickey recounted one of her first questions of me when I reported aboard. "What do you know about fisheries?"

According to Laura, my answer was something to the effect that you look inside the fish for drugs.

She said, "Around here, we measure and count them."

It was a learning experience for me!

Laura brought class to the ship. On *Valiant* and *Vashon*, we had to sweep up magazines and movies if we were expecting someone's mother or children. That was not necessary under Laura. We were way more professional.

I had an interesting moment on my first mixed-gender vessel. My cabin as captain and her cabin as XO were on the level just below the bridge. We shared a bathroom which could be accessed from either side with locking doors. It was the first week, and I didn't even have keys to my room yet.

I have a habit of getting to work early, so I made my way to the cabin, but my door was locked. This was a common practice by the in-port duty officer to lock staterooms after liberty is granted. There was a simple workaround; just go through the XO's stateroom, through the head, to get to my cabin which functioned as both an office and my sleeping quarters.

It was early, so I didn't expect anyone to be in, so without knocking, I swung the door open wide, and there was Laura, with her back to me, changing.

I panicked! I shut the door and ran. I'm not sure she ever knew who it was, but that was a lesson learned that served me well in the future, not only in the service but at home with a wife and daughter. If it's a woman's room always, always knock. When in the bathroom, lock the doors. But when done, don't forget to unlock them!

If you didn't count the Academy cadet commands (and no one does, by the way), the *Key Largo* was my first official Coast Guard command. I finally checked this off the bucket list. After being an XO on the same class vessel, it was exciting, but I also came with some confidence.

Plus, there was a familiar face on board. I couldn't miss that scraggly mustache. Gunners' mate Tom Huffman had been on *Vashon* with me during two years of high adventure in Puerto Rico, and we were ready for round two . . . but he almost missed out.

Not long after reporting aboard the *Key Largo*, we were on patrol and had a short port of call in Beaufort, South Carolina, where they happened to be shooting a movie. The film crew saw Tom and thought he might make a good extra for a battle scene. He said sure, but that he had to check with the ship. As much as I would have loved to see him getting blown up on the big screen, we needed him more for patrol. You may have heard of the movie—*Forrest Gump*.

While the operating area of South Carolina and the northern half of Georgia was certainly different from the wild west of Puerto Rico, the 7th Coast Guard District covered the Caribbean, Florida, Georgia and South Carolina. Group Charleston was so far north of the district, it got the reputation of being the 6th District. There is no sixth district, by the way. The admiral based in Miami had his hands full of migrants and drugs coming up from the south. The group commander in Charleston followed the same philosophy as the 7th District; he believed in being hands-off and allowing junior commanders to command. He told the station master chief and me that we were to run our own commands, and he was there to support us as needed. We rarely saw him. While I liked CDR Pat Boyle, I admittedly liked the independence even more. He served as a great role model on how to command a group.

There were only two patrol boats in Group Charleston, so we rarely had an opportunity to truly relax. When in port, we were almost always in some sort of Bravo recall status, meaning if we were called, we'd have to be underway in two hours.

Living five hours apart, I was on a relationship break with Nye, and we were dating other people.

It really sucks when you finally get the nerve to ask someone out, go to a movie like Forrest Gump, and ten minutes into it, your beeper goes off. That meant you most likely had an urgent search and rescue case that had you sprinting to the boat. When you drop off your irritated date, there is a very low probability of a second encounter. I realized then that first responders, medical professionals, and others providing critical services do this all the time, so hats off to them—and their dates!

Savannah is a beautiful city with majestic oaks covered with flowing gray moss. I bought an aluminum Jon boat with a twenty-horsepower engine and took it in the narrow creeks and tributaries. Positioned right where the tide was running out of a cut in a marsh, a properly cast net filled a playmate cooler full of shrimp in less than an hour. No casting in the lakes though; they were teeming with alligators.

When I was boating, I carried a bag phone. It came in a black leather suitcase and had a spiral antenna that sat on a pedestal. It was my first "cell" phone. I'm sure some of my fellow shrimpers thought I was in the secret service.

We had a great crew, but the operating area, while beautiful, was a bit mundane for my taste. We made many fishery seizures for illegal gear and undersize fish. In fact, ever the competitor, I made a little chart and tracked all the patrol boat cases in the district, and we were leading. We even won a National Marine Fisheries award. But we did screw up once—big time. I'm surprised they didn't withdraw the award.

We were well north in our district and boarded a commercial fishing vessel offshore. The boarding team reported that the vessel had many undersized snappers. We followed protocol and advised the group that we would escort the fishing vessel into Wilmington, North Carolina. This is a huge pain in the butt for the fishermen; I get it. But enforcing fisheries laws is a noble mission none the less and protects species, our food supply, and fishermen's livelihood into the future.

Did I say species? When we got into Wilmington, the boarding team had misidentified the snapper. They were saucer-eyed porgies, a non-regulated fish. The fishing boat was hot, almost as hot as my boss. This was especially bad, given that we had crossed into North Carolina, an entirely different district.

This was like a Coast Guard international incident. It was a human mistake, and cooler heads prevailed. The crew still messed with me by leaving a souvenir saucer-eyed porgy on a plate in my cabin. CDR Boyle delivered a formal apology to the fishermen, and we were out of the doghouse.

From a law enforcement perspective, we also dealt with boating while intoxicated cases, migrants, and yes, even a drug case.

Despite being well out of the main smuggling routes, we managed to get a drug bust assist in a joint operation that netted seventy-four kilos of cocaine and four arrests on 4-6 Oct 1993. Interdicting narcotics at sea had always been my passion, but when a controlled operation works, it really put a hurt on the smuggling community. Our liaising with US Customs paid off when they shared information indicating the M/V *Harold La Borde*, a 460-foot cargo vessel was enroute from Trinidad and Tobago. There was a strong possibility that the ship bound for Savannah had contraband onboard. Specifically, a crew member on board was supposed to pass an unknown quantity of cocaine to two suspects in Savannah. The two Savannah suspects were located and surveilled by Customs.

The day before the ship arrived, Customs agents found a 1 7/8" trailer hitch ball receipt in the suspect's trash, indicating the possibility that the smuggler might try to drop contraband to a small waiting boat. That was the same size hitch I had for my Jon boat. The most logical drop point was the Intercoastal waterway's juncture with the Savannah River, leaving the smuggler six possible escape routes. As the day developed, two other suspect vessels with possible involvement were in the area.

The *Key Largo* and Custom's aircraft located M/V *Harold La Borde* well offshore and shadowed the vessel on its approach to Savannah, ensuring no transfer was made at sea. At 2300 or eleven p.m., *Key Largo* raced ahead in front of the vessel and dropped its small boat with a boarding team at the Savannah tower, a structure a few miles off Savannah's sea buoy in the path of the approaching vessel.

It was a pitch-black night, so we moved out of visual range and monitored the situation by radar. As the suspect vessel passed the tower, our small, unlit boat covertly fell in behind the ship in its blind spot and followed it up the river for the next five hours. Custom's and Station Tybee's small boats blocked any possible escape. As it turned out, the smuggling crewman waited until the vessel moored to transfer thirty-seven kilos of cocaine. Customs agents followed the suspect to a rental storage facility where he stashed the goods. When the ship departed, two other individuals arrived at the storage facility and were arrested by customs agents. The M/V *Harold La Borde* proceeded to Wilmington with the smuggler still onboard and unaware of the interdiction in Savannah.

Upon arrival in Wilmington, the smugglers and the law enforcement agents conducted a similar operation resulting in two more arrests and the seizure of an additional thirty-seven kilos. In all, good cooperation and planning resulted in the seizure of seventy-four kilos, four arrests, and abruptly stopping what was believed to be a recurring smuggling operation.

We were also involved in many search and rescue cases, from disabled vessels to downed aircraft. On one case in February, we had a large, disabled fishing vessel east of the Gulf Stream. Being that far offshore, we were the only vessel available for the mission. The winter seas were very rough, running eight to ten feet. Launching the small boat with its single cable crane was always a challenge, but it seemed even more dangerous that day. Finally, we got the violent swinging under control and the boat to the hip and loaded the boarding team. They made it to the vessel after a ride that would rival any waterpark attraction. Drenched, they assisted with some minor damage control onboard and rigged a sturdy tow. After some nail-biting, we recovered our small boat. With the tow established, we started the one hundred plus mile steam at a blistering four knots.

After heading below, we received the call on the high frequency (long-range) radio that the group was on the line and needed to speak to me personally. My mother was patched through, and she told me that my father had passed away. Perhaps I was a bit vulnerable at that time, having been up for twenty-four hours. I did what I could to comfort my mom, who had to break the news. My crew departed the bridge to give me privacy. After the call, I headed below to the cabin, telling the XO what had happened. This had not been unexpected, but it still hit hard. My father was in his mid-fifties and had been too ill to travel to the change of command for the *Key Largo*. On numerous occasions, he almost died, stretching back to my time on *Valiant*. Just a couple of weeks earlier, thankfully, I had been in Raleigh when he had made the decision to go off life support. He lasted longer than the doctors predicted.

When we got in, my sister ship, the *Metompkin*, like all good Coasties, took over recall status and allowed me to head to Raleigh for the funeral. In a sense, I was sad but also relieved. His quality of life was terrible, and now he would be in a better place. My mom had been a primary caretaker for years, and finally, she would be able to move on.

On my last trip to see him, he presented me with a sealed personal letter and asked me to read it following his passing. Here are excerpts that are applicable for anyone facing the end of their life and wanting to make peace with a loved one. It was long and, apparently, written over several years.

Dear Mark,

By the time you get this letter, you will probably have rendezvoused with Johnny and be in the final days of your "Puerto Rico Junket." To us, it doesn't seem possible that you have completed four years of "Sea Duty" already. As you complete all your duties for the last time and pass on responsibilities to your successor, it must give you a good feeling of accomplishment. Your distinguished record will not be easily emulated. It is time to come home, rest, bask in the glory of your accomplishments, and begin work on the next important chapter in your life.

New beginnings are stimulating and help develop the best that we can give. In retrospect, I wish I had experienced more of them in my life. If carefully chosen, they can help us do more with our lives.

You and Johnny have almost perfected the strategy of seeking out responsibilities and growing with the opportunities that present themselves to you. In fact, you two guys have been quite effective in teaching me more than you probably know. As mom & I have reviewed your progress through scouting, sports, and the Academy, we have witnessed a beautiful transformation taking place. The only way I could be more proud of your growth and leadership development is to have been able to contribute more directly myself

As we look at the other young men your age, it seems that there are at least two characteristics that are strongly influencing your life that may be less present in theirs. One is a sense of confidence, accomplished by a strong drive to succeed. When this energy isn't focused on a career or goal, an individual can appear to be unmotivated or intend to just tread H20. The other quality you have demonstrated is the willingness to leave home and take charge of your life. It seems only logical that the opportunity to grow, to become independent and responsible for your life would be enhanced by leaving the security of the nest and testing your own wings. Even if they get clipped on occasion, just think how much can be learned in the process.

Well, now I've done it. You know my illness has greatly cut into my energy level and involvement with people. But while curbing my activities, it has also given me cause to reflect, read, and almost become philosophical. As I share insights or comments with you, don't hesitate to challenge me when you sense that perception is way off base.

For the first time, after almost thirty years of marriage, your mother and I are beginning to really communicate with each other. It is sad for it to have taken so long to get to know the inner thoughts of the person with whom you've chosen to share your life, but it is better than never. By personality and by upbringing, it is not surprising that we have found it difficult to be closer to each other. As we read and talk, we are finally beginning to capture the closeness that has eluded us in the past.

Why am I telling you this? As I continue to wrestle with my health problems, it is easy to see evidence of my losing ground. Considering the odds and the circumstances, this is certainly expected. While I am not even close to throwing in the towel, there is a personal agenda that I need to complete.

High on this list of priorities is the desire to visit privately (which means including your mother) with each one of our sons. The purpose is to handle some unfinished business; complete my parenting; discard old unnecessary baggage, tie up loose ends, resolve old conflicts, seek forgiveness for my shortcomings.

I have no idea how receptive you guys may be to this. It is a big and important issue for me. For many years I wanted to have father-son talk with Paw Paw. On many occasions, I tried, but failed. We were not on the same wavelength. He was such a private person that he could never reveal his inner most feelings or emotions, and to a great extent, his passing meant I lost the chance to learn about his roots. Dad was a major force in the shaping of the lives of my brothers, sister and me. While there was no doubt about the depth of his love, it was also clear that he was close to being a child abuser. The behavior of his children clearly provides such a chronicle.

Part of becoming a strong individual is having come to terms with one's past. This was a glaring deficit in my life. In these, my twilight years, I am determined to find and destroy (almost a Coast Guard mission) some of the skeletons that have been hiding in my personal closet. And since I don't wish to pass a family pattern on to my sons, I do want to follow through on offering each of you the opportunity to rid yourself of excess baggage. Each of you boys has taught me much, but, Mark, you have given me the best opportunity to be a good father and I have failed to respond. I did not measure up to my ideal of being a good father to you.

To see what you have accomplished with your life makes me profoundly happy for you. It shows the strength of the human spirit. If a person believes in himself enough, there is almost nothing that cannot be accomplished. If only I would have shown you more love and support when you were younger, how much happier would you have been? Perhaps we will never know. We only live our lives once; we can't have it both ways.

When we are together, I hope we can review the past. While I love you greatly, I will not seek forgiveness for the things I've done and not done—sins of commission or omission. Please don't misunderstand me. Overall, I see myself as a loving, caring father. There is almost nothing I wouldn't do for anyone of you guys. But on the negative side, I don't want my weaknesses to generate long-term effects for you. Sometimes, in talking through frustrations, problems, and memories, we purge ourselves of some of the "bad" that they represent. In seeking forgiveness for my shortcomings, I would hope that a degree of cleansing might occur that would release you from some of the shackles of the past. At any rate, that is where I am at right now.

We are eager to have you with us. The reservations are made for our trip to Springmaid Mountain June 11-14. Chip is chomping at the bit to see you. Another 4-runner has slipped through our hands. Sparky is waiting to sleep with you.

We will fish, go horseback riding, and hike. You and Scott will be able to canoe and go rafting on the little mountain river. We will eat well.

It is wonderful that both you and Johnny will soon be near again. We are ecstatic about that.

Take good care of yourself. Let us know what flight to meet on 1 June.

Love, Dad

PS: We want to thank you for the way you have kept in touch by mail and by phone during the 8 years of separation from you. You get an A+ on quality and consistency. With luck, we might eke out a C- by comparison. See you soon.

PSS: Moms mothers day flowers.

I am thankful that I indeed had that conversation.

Upon my return to *Key Largo*, we resumed recall status. Nye had come down during this crazy period. We were done with the break and discussed marriage. I was glad she had met my father before he passed.

One thing about these Coast Guard jobs, you don't have to wait long before the command center calls. This time there was a fishing vessel in the Gulf Stream that had found something other than undersize saucer-eyed porgies. They had come alongside a small, adrift, ski boat with twenty-two people appearing to be Haitians onboard. We had never seen Haitians this far north.

The fishing vessel reported the occupants to be in bad shape, and a two-year-old child had already died. They had had no food or water, apparently for days.

We had missions in the "claw," Port Au Prince, and Cape Haitien in addition to logistics runs to Guantanamo Bay, Cuba.

I'm not sure if we beat the time to get underway set by the *Vashon* for the hijack case, but it was very close. We sped out to the distressed vessel as the fishing vessel stood by to assist. Had that fishing vessel not investigated, these refugees would not be alive today. The fishermen truly were the heroes.

This was a migrant smuggling operation gone wrong. Two Bahamian smugglers were trying to move these people to Florida, but their engine died. They refused to call any passing vessels for help and hoarded minimal food and water rations. When we came alongside, the travelers had immediate medical issues. We quickly got them aboard and placed the child in a body bag. The policy back then was not to take migrants into the skin of the ship due to possible disease and security issues. The other policy was not to land them in the United States. We had to take them to a larger cutter for repatriation. I knew the policy well from my time on *Valiant*. The command center in Miami followed the checklist and directed us to run south to intercept a larger cutter. I said negative.

We already had one fatality, they were badly dehydrated, and if we stayed out, they would perhaps get more seasick with an oncoming storm building to the west. They could not afford to lose any more vital fluids. Lee Petty, *Key Largo*'s new XO who had relieved Laura in the normal crew two year rotation cycle, was way ahead of me and with the crew had brought the mother of the deceased child inside the ship's skin. She was showing signs of shock and needed immediate medical attention. We didn't have a doctor or paramedic on board.

Petty Officer Shaun Edwards was our EMT and our cook. The cook had access to knives and conveniently worked on the mess deck, which doubled as an operating table. Perhaps that's why, historically, the ship's EMT is the cook. The rest of the crew had at least some basic first aid training. Thankfully for the Haitian woman, no surgery was required.

As the CO, there are always tradeoffs in these search and rescue situations, and many times there are life or death consequences. In this case, we could have slowed the ship and run away from the coast to find the best ride and wind direction to execute a helicopter hoist. That would have gotten the mother to the hospital slightly faster but delayed medical attention for the rest and placed us in a worst weather situation entering port. I opted to forego the dangerous helicopter hoist evolution and run at best speed into the port with all survivors aboard.

The weather was building quickly. The distant skies over Savannah were dark and filled with lightning. It was a race to get into Tybee Island. The good news was we were the jet fighter of the cutter fleet. Calling ahead, we had ambulances at the dock. While driving from the flying bridge, one of the two Bahamians began to tell me this crazy story about fishing which basically implicated him as the smuggler. Focused on getting in and no time to read him his rights, I just listened. Once we made it, I transcribed my notes and passed them to an immigration officer.

Later, those admissions, along with my testimony and that of the deceased child's mother, were damning to the smuggler's defense. They were convicted of smuggling and manslaughter.

After the ambulances sped off, and I finished a TV interview, I went home, showered, and hit a local restaurant for dinner. Above the bartender was a big TV. There I was, being interviewed with one of those now frequently seen pandemic masks draped around my neck. Drenched from rain, I described the mission, what I knew, and that the fishing vessel saved the day. Several patrons recognized me and bought me drinks that night. I felt great having been part of the rescue but regretted not being able to save the toddler.

That wasn't *Key Largo*'s last encounter with Haitians; in fact, that was just foreshadowing. In the summer of 1994, all but four of our sixteen-person crew rotated off. These were only two-year assignments for a reason as they were physically demanding on the body! Laura Dickey moved on and lead a distinguished career and is now a rear admiral. I could not have gotten a better replacement at XO then Lee Petty.[1] Shaun Edwards, the cook, went to OCS and made commander. He's now in command of the Gulf Strike Team. Probably good he didn't go into medicine.

The news and message traffic coming out of Miami was that Haiti's political situation was deteriorating rapidly. There was an expectation that thousands upon thousands would take to the waters in makeshift craft to try to escape the failing state. We were once again ramping up but with a very green crew.

---

[1] Lee Petty went on to command the Patrol Forces in the Persian Gulf and became a district chief of staff

Lessons Learned:

It's better to be a kingmaker than a king, and a small unit of twenty or less is probably the most fun.

Not being located with your boss has its advantages.

If you want to make a drug bust outside of traditional smuggling routes, you have to work for it.

A Jon boat with a reliable twenty-horsepower engine is a great starter boat. But always have a paddle aboard.

If you've got money, get an electric motor and carry a "bag" phone.

Most guys I know don't communicate well on emotional subjects. Find a way. Spend time with aging parents. It's important!

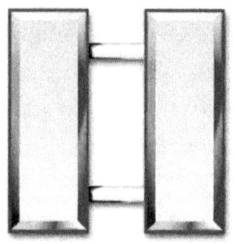

Chapter 16

Haiti Has Fallen

OPERATION ABLE MANNER WAS in response to a humanitarian crisis in Haiti. The Coast Guard forward-deployed all available cutters to deter this dangerous sea migration, meaning we left the confines of our operating area off South Carolina and Georgia for points south. The United States, as part of a United Nations effort, essentially took control of the Haitian side of Hispaniola as the island nation was spiraling into chaos. The *Key Largo* was tasked to deploy for a month and stage out of Guantanamo Bay, Cuba, which lies just to the west of Haiti.

We made it to Miami and refueled, but there was a major storm between us and Haiti.

The cutter that was already off the coast of Haiti was very anxious to have us in the theater. One of our favorite movies that we played all the time on *Key Largo* was the original *Top Gun*. This situation was like being Maverick on the carrier, ready as back up, listening to our sister cutter in a world of hurt. The cutter off Haiti was like Iceman—surrounded and outnumbered in need of immediate back up.

Pressured, we got underway with several other patrol boats despite the weather. Shortly after leaving the protection of Miami's jetties, we hit over fifteen-foot seas. One wave crashed over the bow and broke the chains holding our new 25mm gun. It didn't break the mount, but it spun all the way around. That was enough for me to decide to turn around. We waited another twelve hours until the seas abated.

Once in the old Bahamas Channel, we made a good time. At one point, while in international waters north of Cuba, a Cuban Border

Guard boat known to harass passing cutters attempted to intercept us. We left him in the dust.

When we arrived in Guantanamo Bay, we embarked two marines to augment our security and an Army Special Forces team. Our mission was to insert that team using our small boat into rivers or coastal areas in the "claw" of Haiti to recover both weapons and prisoners from abandoned jails. The thought was to hopefully minimize retaliation against the still-confined convicts and reduce the number of weapons on the street.

When I was told we'd be taking on Army Special Forces guys, I was expecting someone that looked like the actors Chuck Norris or Jason Statham, but the real SF guys were pretty laid back. In fact, it was difficult breaking them free from our mess deck, where they drained our supply of ice cream sandwiches. When they eventually departed, they went heavy with body armor, helmets and machine guns. Our sea painter, which is the line that tows the small boat alongside the cutter when launching, broke under the heavy, Army, landing party. No harm done, but we rigged a heavier rope to accommodate.

These Army guys were a good group, as were our two young Marines who also augmented the crew for extra security onboard. This force was sent up rivers where they would disembark and seize numerous rifles, pistols and grenades, primarily at Haitian police and military installations. With the country's deteriorating conditions, this was a precaution in case these facilities were overrun by unfriendly elements.

When not doing Army insertions, we anchored in the port of Port-au-Prince. At the two-week mark, we shifted to the north coast off the port of Cap-Haitien, which was much tighter to get into. Initially, we stayed at anchor, which afforded better security than tying up.

Key Largo's schedule was on patrol for roughly five days in and around Haiti. At day five, especially with Army soldiers and Marines aboard, we'd have to head back to Guantanamo Bay to refuel and pick up groceries. We also gathered Haitians from the water or rafts on the way and carried them back to Cuba where another US Army unit had erected a tent city on the eastern entrance of the port.

There are a lot of people who think those in the military are soulless brutes who just want an excuse to kill someone. Well, those misinformed people need to leave their institutions of higher learning and come to the frontlines. We picked up a group of Haitians, one of whom was a young man, perhaps in his late teens or early twenties. He was paralyzed from the waist down and stricken with aids. Despite being in very rough shape, he was obviously a kind young man. Two of our seamen, probably about the same age, volunteered their valuable off-watch time to care for him throughout the night, transiting to the Guantanamo Bay detention camp, otherwise known as GTMO.

Our home port away from Savannah was rapidly becoming GTMO. While on the communist island nation of Cuba, the United States had secured the port during the Spanish-American War over one hundred years earlier. It's technically United States soil. I had trained in GTMO as an ensign aboard the cutter *Valiant*, so I knew the lay of the land. The naval compound was a little slice of freedom surrounded by fences, guard posts and minefields.

If you've seen the movie *A Few Good Men*, Colonel Jessup, played by Jack Nicholson, does a good job of describing the life and mission of the marines on the base. It has to be self-contained, from aircraft runways to power stations to desalination plants. Essentially, it's a fortified foothold in an unfriendly country. Eventually, it housed some of the worst terrorists following 9/11.

Time off was brief, and there wasn't a lot to do. Recreation consisted of a base exchange, an outdoor free movie theater and only one good beach called Cable Beach. Navy, Marine, or Coast Guard crews were familiar with this primary recreation site in an otherwise austere plot of land. *Key Largo* was tied up at the pier, and we were not supposed to be called upon for the next twenty-four hours. The crew hiked up the hill to do laundry and get food stores, while some went to the beach.

I was still on board when I received an urgent call from the captain of the 378-foot Coast Guard cutter anchored in Guantanamo Bay. He ordered us to get underway immediately.

We watched well over one hundred Haitians jump from the cliffs into the bay near their tent city on the east side of the harbor. Our interpreters said those Haitians who had jumped from the cliff were trying to swim across the harbor to the west side. They thought, incorrectly, that the United States controlled the east side, and the Cubans were in control of the west side of the mouth of the harbor. It was actually all US property. In fact, even if the Haitian swimmers had made it across, alluded the small boats and land forces, they would still have had to traverse a minefield and awaiting Cuban border guards.

That said, there were numerous refugees in the water without life jackets, and the large, anchored cutter was rapidly getting overwhelmed. They had launched their small boats. Our small boat sped off to start the rescue. That didn't cut it either. They needed the *Key Largo* out there too!

After taking a quick look around, knowing over half the crew was at Cable Beach or on the small boat, only four people were left, including myself and a Marine, to get the 110-foot ship underway. A bit unheard of (and also not recommended except in urgent situations), we got the ship out there with the skeleton crew, linked up with the small boats and started bringing in exhausted migrants. One Haitian man remained defiant, though. Our young marine with me on the flying bridge aimed his rifle at the man treading water.

"No . . . he's not the enemy!" I yelled.

AIDS was widespread among the refugees, so we had to avoid getting cut or bitten by a hostile swimmer. Eventually an interpreter talked him down, and he cooperated. Amazingly there was no loss of life during this rescue.

The final, four-day sortie of the Haitian mission was to relieve a fellow, 110-foot, patrol boat currently in the northern port of Cap-Haitien. The smaller patrol boats went into the ports and rivers due to the water depth, while the larger cutters remained offshore. During my relief of my sister ship, I was told there was an Army insertion mission planned for the north coast the next day. It was a logistics supply deal, nothing exciting. When I relayed to our new fleet commander aboard the 270-foot cutter Bear—ironically the same cutter we worked with during the St. Croix mission—I was told to stand down and cancel the joint mission.

That was not going to sit well with our Army partners who had been planning it for a week. I did a "but, sir." Apparently, the CO didn't appreciate that and ordered us to get underway and rendezvous offshore for a face-to-face meeting. As we closed on the cutter Bear, the crew hummed the Star Wars soundtrack from when the good guys were getting sucked into the Death Star.

The XO met me at the fantail and escorted me to the captain's cabin. It felt like I was being led to the gallows. I must say, he wasn't my favorite *Bear* skipper, but at least he understood more about our relationship with the Army, the mission that had be agreed upon by my processor, and *Key Largo*'s need to move offshore periodically to make our own freshwater with our reverse osmosis water makers vice take on potentially contaminated water in the Haitian port.

When *Key Largo* returned to Cap-Haitien, we tied up to the Army landing craft and did operations with them. Because we definitely couldn't take on water in Haiti (I did not want a sequel to the India milkshake), we had been rationing water. The conversation face to face cleared the way to periodically move offshore to make fresh water.

I went ashore with our Army partners and even drove through the port area in one of their Humvees. I saw a dead body on the side of the road, but people just walked by it. In Port-a-Prince, where UN troops were housed, I witnessed a UN peacekeeping soldier in a bunk bed take a crap in his hand, toss it on the ground, and go back to sleep. Learning point: that's why it's taboo in some countries to shake or point with the left hand. You're welcome for that image.

I knew LCDR Ed Daniels, the CO of the buoy tender which was part of our little task force assigned to Haiti's mission. His ship was also assigned under the *Bear* task force. We established a backchannel to communicate. On one very nice day in the claw, we rendezvoused far away from shore and held a swim call. We even used their big buoy crane as a rope swing. We were having a blast until Bear's helicopter overflew us!

It was almost the end of the month's deployment, so we headed back to GTMO for a quick overnight refuel and prep for the long trip back to Savannah. The crew was tired but happy with all we had accomplished. Most had already called families and girlfriends to let them know we were finally on the way home.

Bunks on the *Key Largo* were three high in some places. There was no washer or dryer, so we joked about putting restraining orders on each other when we got back. Spirits were high. Someone proposed that we head out to a grassy field on GTMO and play a game of football to get a little exercise before the long transit home. Except for the in-port watch, I think most of the crew went, including me. It was perhaps a half-mile walk. Once we got there, the clouds rolled in, and finally, there was a deluge. GTMO looks like a desert, so when the rain dumped, it didn't seep into the baked ground. We kept playing. If you have ever seen mud wrestling, well, that's what it looked like. I don't think anyone knew the score at the end. When we finally called it, the only thing white was our eyes and grinning teeth. We passed two truckloads of sailors or marines heading back to the boat, and they looked at us and started cheering.

The next morning as we prepared to get underway, the bridge radio came to life. It was the 7th District Command Center. We were not going home. Cuba was now in chaos, and thousands of Cuban refugees were massing on the northern border, preparing to take to rafts to cross the treacherous seventy-mile gap to Florida. We were to make the best speed to Key West and would be stationed off the hottest spot, twelve miles north of Havana. I broke the news to the crew, and they were less than thrilled. The only positive was we had to go that way to get home anyway.

Lessons Learned:

Six-week deployment on a small boat is a long time.

Getting *really* dirty can be a lot of fun.

Compassion for those less fortunate can be heartwarming.

Being the bearer of bad news is a command responsibility.

If you are going to goof off, make sure the boss doesn't have a helicopter.

Sometimes you have to stand your ground to a senior leader. Just make sure you are in the right and your ducks are in a row.

Chapter 17

Can't Go Home Yet

SIMILAR TO THE POVERTY-DRIVEN exodus in Haiti, Cuba had fallen into an economic crisis because of the recent collapse of the Soviet Union and the ongoing American embargo. Many Cuban citizens had stolen boats while others participated in a raft exodus or Balsero crisis.

More than 35,000 attempted to make it to the United States on makeshift rafts. The heavy vocal Cuban-American population of southern Florida had family members trying to flee oppression, which made the whole situation a red hot political and life-threatening event. *Key Largo*'s assignment was to take up a station between Havana and the Florida Keys.

After the transit north, we arrived in Key West and found the basin to be a parking lot for 110-foot patrol boats. Operation Able Vigil had officially started. I counted twenty patrol boats who had arrived from the Caribbean, east, and gulf coasts. They were breasted out like we had done at the squadron in Puerto Rico.

I knew probably three-quarters of the other captains from the Academy or previous tours. One of the first things we did was get a huge load of migrant supplies that included life vests, blankets, rice, beans, water, basic personal hygiene products and medical equipment. This was the same type kit we had for the Haitians.

Haiti was a little challenging because the migrants spoke Creole, and those interpreters were hard to come by. In a lot of cases, we resorted to hand signals. But the Cubans, of course, spoke Spanish, so we had an endless supply of bilingual Spanish-speaking Coasties. We were also now quite comfortable with doing migrant operations. We no longer had the green crew we had sailed with a month earlier. Everyone onboard knew exactly what to do and when to do it—almost everyone.

Nye and I had gotten back together after a nice weekend at Hilton Head, South Carolina. This was just before sailing, and in fact, she was in my bachelor house in Hampton sanding and painting out the garage when she tried to reach me in Key West. In a very rare opportunity to leave the ship, the crew and I went out to Duval Street for a little R&R. No bar fights this time, mind you. But when Nye reached the ship's phone, the cook, who was the in-port duty officer, thought it wise to come up with a cover story since I was out on the town.

Later that night, Nye and I connected. The cover story didn't exactly match up with the story I told. Between that and the fact that I was goofing off while she was sanding didn't sit well. I bought her some nice t-shirts to go with the Colombian emerald necklace I had previously risked my life for, and we were back on solid ground.

That night off was the last of the R&R. We were so busy picking up rafters nonstop that resupply in Key West focused purely on refueling, grabbing groceries and migrant supplies, and immediately heading back out. We hardly noticed when the deployment went into its sixth week, rivalling my longest deployment on the much larger cutter *Valiant*.

In addition to the patrol boats, larger cutters and naval ships were also present in the Straits of Florida. It was the biggest collection of ships I had ever experienced. It needed to be. The numbers of migrants dwarfed what we had seen coming out of Haiti, where migrants had used repurposed boats, whether they were sailboats, coastal freighters, or fishing boats, to make their escape. But here, if it could be floated, Cubans were on it. Fidel Castro had lost control, and thousands wanted out of communist Cuba!

The *Key Largo* drew the hot zone off Havana. The strategy was simple: Form a ring of steel and pick up the rafters. Don't let the rafters get through.

If we miss them, they could easily get caught in the Gulf Stream and swept north to their death. I learned that during our Haitian case off the coast near Savannah.

Being underway was like participating in a Fourth of July boat parade. There were rafts as far as we could see. It was almost easier for us to drift and have them paddle to us. We were not allowed to enter the twelve-nautical-mile territorial sea of Cuba, but we could hug the line, and the mountains of Cuba gave us a great radar return to fix our position.

There was also a national policy for handling Cuban refugees. If they got *feet dry* in the United States, they would undergo a lengthy procedure to determine if they would be granted asylum. With the numbers pouring out of Cuba, our job was to pick them up quickly and deliver them to the larger Navy holding platforms. Eventually the large navy ships also got into the pick-up operation.

At one point, we had probably sixty migrants on our deck, and the lookout reported a black smoke cloud off in the distance. It was significant and coming from one of the Navy holding ships. Then over the VHF radio came a broken message requesting assistance. At sea, the very last words you want to hear are "we're on fire!"

The navy ship had come across a raft that had used fifty-five-gallon barrels as its flotation device. Our process was to get the migrants off the raft, then sink the raft or at least mark it so someone coming across it knew the people had been safely removed and weren't treading water somewhere. These Navy ships weren't very maneuverable. After safely removing the migrants, they lit the raft on fire to destroy it. When they did, the gasoline residue in the barrels caught fire, sending flames high into the air.

The flames caught the side of the Navy ship on fire. The smoke was so thick that the ship's engines completely shut off when smoke was sucked into the air intakes. That left them essentially alongside a burning raft, unable to move. The fire put a fat, black scar on the ship's side.

As the closest vessel, the *Key Largo* raced to assist, even using the migrants to help us set up the fire hoses on the bow. By the time we got on the scene, the fire was out, but the damage was done. It took the Navy ship several days to get back fully operational, but they didn't have time to paint over their black stripe.

Eventually, the larger ships repatriated the migrants directly to Cuba or took them to Guantanamo Bay for further screening. While we hovered at the twelve-mile line, a boat full of refugees paddled their butts off toward us. Castro had lost most of his control but, unfortunately, not all. Just behind the paddlers, a Cuban Border Guard boat chased them down.

We did everything to encourage them to speed up their strokes, as it was close. At one point, one of the Navy ships said the Cubans had the *Key Largo* locked up on their fire control radar. We didn't have those sensors, so I encouraged our Navy friends to get a bit closer to us.

The migrants thankfully made it, and the border guards broke pursuit.

On one of our brief refueling stops, the *Key Largo* was selected to host an international news crew. They were to stay clear of our operations but certainly wanted to get close up video of Coasties pulling desperate Cubans from the water.

Not all rafts are created equal. Here's a couple of ingenious Cuban transports:

I think the Cuban migrants could have made my Camaro a boat.
(photo courtesy of DVIDS)

Cuban migrant trucks don't need roads.
(photo courtesy of DVIDS)

With *Key Largo*'s hull being about the thickness of a beer can, we really had to be careful not to hit anything. One pitch-black night, I had positioned myself on the flying bridge, propped up, using the night vision scope. Having a little background light helps to discriminate targets on the flickering green screen. I saw a raft off the bow. Thankfully we were just drifting.

When I started to twist the engines to recover those on the raft, I heard screaming off the stern. It was a single man in dark clothes on a small black inner tube some sixteen miles from shore. We safely recovered them all, no thanks to the international press crew with their blinding TV camera lights.

These refugees were different from what we had experienced the weeks earlier off Haiti. Among those brought aboard this time were doctors, lawyers, and even a government secretary. They were entire families, including pets, which would occasionally relieve themselves on *Key Largo*'s freshly painted nonskid. Even the deck force guys charged with cleaning and painting didn't seem to mind. They saw how important the dogs were to these families, especially those with kids. It was tight quarters, though, and all the migrants had to remain topside. The best way to ensure the situation didn't get out of control was to treat everyone warmly. They had been effectively traumatized.

We presented the kids with small mementos or candy. We also tried to pick out doctors or nurses to help identify and treat any immediate medical concerns such as infants and pregnant women.

Our mission was in the news extensively, so when we were finally released, we got a bit of a hero's welcome

The majority of the Cuban migrants were on rafts like this. (photo courtesy of DVIDS)

returning to Savannah. Operations Able Manner and Able Vigil collectively picked up 63,000 migrants.

Sixteen guys living in a very small space under high stress conditions meant one thing: we all wanted space from our shipmates. It had been an epic adventure, but I was nearing the end of my time on *Key Largo*.

It was time for me to move on, and I got some great news. My classmate and good friend Mike McAllister was scheduled to relieve me. The *Key Largo* would be in great hands, and I was given the job as the assistant chief of Search and Rescue for the combined Fifth District and Atlantic Area in Portsmouth, Virginia. One part of the job was to serve as the Atlantic Area Patrol Boat Manager overseeing sixty vessels like the *Key Largo*.

As an added bonus, I had a freshly painted garage in Hampton, and I could move back into my house with Nye.

Lessons Learned:

The Cuban people are very resourceful. I suspect they could have made my Camaro a boat.

Close distance on a missile boat if you are armed only with guns.

If it's pitch black out in a sea of rafters, better off drifting than running at clutch.

The limit of refugees on a 110-foot patrol boat is 150. But in a pinch, I saw one with 300.

When destroying rafts by fire, make sure you have a good stand-off distance.

Facing a mass exodus by water, the credible threat of repatriation can be a lifesaving deterrent.

Make sure your cook is fully briefed on your love life.

Despite how you think you might look in comparison, given a choice, it is always better to relieve and be relieved by great officers.

This was a common site. 110' patrol boats like Key Largo loaded with migrants being delivered to large Navy holding ships. (photo courtesy of DVIDS)

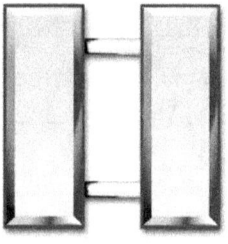

Chapter 18

Powerful Carpool, Kazakhstan and Turkey

I WON'T KID YOU; I wasn't necessarily thrilled about going to what I thought would be a boring staff tour. Hanging out with a bunch of senior staff officers and pushing paper didn't sound like a lot of fun. I would not, nor would my bosses to this point, ever categorize me as a good staff officer. In fact, I didn't even read my orders carefully for the first week. I thought it would be three years behind a desk, and then I could go back to the action. Nope, I read it wrong; it was a four-year assignment in the Portsmouth, Virginia Federal Building.

The Coast Guard is constantly getting its budget squeezed, and while I was reporting into the Search and Rescue Branch for the Fifth District, which covered the Mid-Atlantic, there was a merger afoot. The five east coast districts were commanded by one or two-star admirals who worked for the three-star Atlantic Area commander. It just so happens that the Atlantic commander and Fifth District commander were both in the same building in Portsmouth. To make matters even more interesting, the Coast Guard decided to combine the two commands, deleting the need for one admiral and staff.

When I arrived, there were eight people to manage the Fifth District Search and Rescue program and fifty-two units. Then we took on the Atlantic Area duties and dropped the staff from eight to two. I was one of the lucky two, but at least I was working for a great pilot named CDR Bruce Frail and his cutterman boss, Captain Mike Ragsdale. This was right around the time of the new DC buzz phrases of "reinventing government and streamlining." Doing more with less sounded good on paper and to the tax paying public.

146

As a former patrol boat skipper, one of my primary jobs was to manage the sixty Atlantic Area homeported patrol boats. That included conducting a formal homeporting study to improve efficiency. Here's a learning point for a young staff lieutenant: if you're told by the big boss (Atlantic Area Commander) to come up with a study that shows we need to permanently move patrol boats from New England to Puerto Rico, there will be winners and losers. When briefing the admiral in charge of New England, you need to demonstrate you have evaluated both pros and cons of such a move. Superb salesmanship only works well when they get something in return. We eventually moved those patrol boats south, but those briefs to the Coast Guard and Congressional leadership in New England were painful.

In addition to that bait and switch by my assignment officer, by this time in my career, I was considered a special forces guy. We had the Atlanta Olympics on the horizon with water events in and around Savannah, so I was selected—due to the recent patrol boat assignment along with IMLETT skills—to be on the Battle Roster Port Security Unit or PSU 302. I had a couple of buddies also selected: Joe Rodriguez from DIAT and Keith Smith who I got to know briefly in Puerto Rico on a drug mission.

The commander of the unit was a real friendly captain named John Cook. He was also assigned to the district and ran the Aids to Navigation program. When not doing battle roster port security missions, I supported Captain Cook who ran the Command Screening Panel for enlisted officers in charge. Those talented enlisted candidates, if qualified, commanded stations, patrol boats, and aids to navigation teams. One of the greatest features of the Coast Guard is that our junior folks are given enormous responsibility. Rarely ever do they disappoint.

Back to the battle roster selection. We went to Camp Perry, Ohio, for basic infantry training. I qualified in things like a grenade launcher. That would have been helpful in El Salvador but better late than never.

In the end, I didn't deploy to Atlanta for the Olympics but rather was placed in the fledgling Port Security Unit 305 based in Hampton Roads at the Army base Fort Eustis. PSU 305 was a reserve unit with a small active-duty support contingent. To be clear, personally, this was a big commitment: one weekend a month and a two-week deployment in the summer, just as a reservist did but without the extra pay. To complicate an already demanding work schedule, I had just gotten married and had a kid on the way. Making the next rank without a master's degree was pretty hard, so I had gone back to school at night too.

After three classes at Old Dominion University, I dropped out, rationalizing being a parent and husband trumps doing papers all weekend. Doing more with less had its limitations! During this tour, my saving grace was probably connecting with a few cool guys for the daily ride to the Portsmouth Federal Building. Dean Lee was all about small boats and search and rescue. He had a booming deep voice and tons of funny stories. He had been a search and rescue station CO in Atlantic City and was an Officer Candidate School (OCS) graduate (versus the Academy).

The other guy was my battle roster PSU shipmate, Keith Smith. He was prior enlisted and then went to OCS early in his career. If you want to blame anyone, Keith was the one who invited me into the carpool. He was the CO of the Tactical Law Enforcement Team-North that resided in the federal building's basement. There was also Chief Ed Guy, who was on the team with Keith. He had a corvette!

Then there was Jim Obemesser. Jim was a marine safety type. He was the odd man out in the carpool. Marine safety made up about half the officer corps and was shore-based, essentially consisting of inspectors and regulators. He was laid back and a great guy who cared about everyone. We all loved him, but boy, Dean and Jim went round and round on what career paths were more important or exciting. Dean Lee used to say he'd rather dig his eye out with a spoon than read regulations.

One of the cars had a fancy, high-tech, fold-down TV screen that played DVDs. Being a bunch of newlywed guys, we watched episodes of *Jack Ass*.

I received a portable thermostat for Christmas. We sealed up the car to see if we could get the inside temperature to break 130 degrees, our own portable sauna. Then there was the daring rescue with Ed of a poor dog with a broken leg stuck on the bridge. Many cars were too busy to help and pressed on, but not us. We closed down traffic and corralled him, and then Ed drove the bleeding dog to the vet . . . in a 'vet.

I was happy to take it all in and learn from these great leaders, but it still didn't keep us from complaining privately about the Portsmouth Federal Building. It was like a prison camp for junior officers. Fast forward twenty years. Jim had succumbed to cancer, but he was so revered that they named the main meeting room in his honor. My carmates had something to do with that. Dean Lee would get his third star and serve as the Atlantic Area Commander and Keith Smith his second star and serve as the Fifth District Commander.

Being assigned to the 140-person Reserve Port Security unit for two years was an eye opener for me. Many of the reservists were police officers or government workers, but some were students, and others were car salesmen. Many were older, and there were even Vietnam War veterans. They were talented leaders who had joined the PSU, not to manage yet another staff, but because they were paid to drive fast boats with machine guns. To many, it was not work. To others, it was a critical source of income and benefits for their families, especially if the economy tanked and they were laid off from their primary job.

We had six, center-console Boston Whalers that carried mounted fifty caliber and M240 machine guns. We also had a shore or land force that I oversaw, in addition to being involved with the boats and operations.

Port Security Unit on the water in Turkey
(photo courtesy of DVIDS)

It was a relatively new unit designed to be self-contained and to deploy overseas to protect military shipments arriving in a foreign theater. This unit housed some of the few Coast Guardsmen who deployed to the Gulf War. The PSU would be embedded under a Navy-led harbor defense command when deployed.

When I was on the international training team, I was by myself or with a team of up to four people. We worked closely with the embassy and the local country's military or police. We generally stayed in nice

Port Security Units built gun emplacements ashore (photo courtesy of DVIDS)

hotels, but not always.

This particular deployment was a defense mission. We were in camouflage, hiked in, constructed a tent city, set up defensive positions, and built our own floating pier. We had thirty people in a tent. It was more like a Boy Scout trip than the Coast Guard I was used to. It was new and, frankly, fun. One thing you figure out quickly when deployed on these missions is to sleep when you can. You never know when you'll have to be awake for long periods of time.

I was selected for LCDR but still wore LT when we deployed for two weeks to Marmaris, Turkey. For the record, Turkey is Savannah-hot in the summer!

Many restrictions were in place to allow us to camp on the Turkish naval base. No photography allowed, and water was rationed, so you could only shower every other day. After two weeks of Meals Ready to Eat, the novelty was starting to wear off. The toilets, well, let's just say that was an adventure. There was a containment system; that's a nice way to say trough. The stipulation was it couldn't accept the paper; there was a trash can for that. That meant we got good at quickly taking off and replacing the lid.

Turkey was an important NATO ally and the gateway between Asia and Europe. Istanbul—formerly Constantinople—was the eastern capital of the Roman Empire (I knew those four years of Latin would pay off). Now a Muslim country, the Turks demonstrated a delicate dance by having us onboard.

After a series of sweat-drenched battle problems, we finally got to shed the helmets and enjoy a little time off. One of the guys rented a Samurai jeep. My fellow junior officers Steve Pope, Roger Bullock, John Settle and I ventured into the countryside.

Steve was a captain in the Virginia Marine Police. John Settle was in

Virginia's Department of Environmental Quality and a state on-scene pollution response leader; Roger was an Army Corps of Engineers civilian. Having this relationship with these agency leaders helped tremendously when I took over the command years in Virginia.

But back to the Samurai jeep. We decided to drive as far west as possible to go swimming and, hopefully, see the Greek Isles in the distance. I didn't realize it, but Turkey and Greece had chilly relations at the time, so asking for details about their neighboring country drew less than helpful directions. For the open jeep ride, I had put on suntan lotion that apparently was an aphrodisiac for bees. It was almost a flashback to the

Our jeep could outrun the bees.

yellow jacket experience in Boy Scouts. We had to keep the speed up on the bumpy cobblestone road until finally getting to a suitable beach where I washed off. My buddies continued to remind me of this organ-jarring trip years later.

Besides the bees, Roman ruins dotted the landscape along our route. We got to the end, but our quest was less than satisfying as we could not see the Greek Isles. We did find a neat cave that was very dark (perhaps we had seen too many Indiana Jones movies). We headed in, feeling the walls as we moved forward. It was almost pitch black, but I detected a gap in the floor just ahead. Before trying to step over it, with our combined 160 years of life experience, we decided to light a match and drop it to see the depth of the pit. It fell and fell and then fell some more. We never saw the bottom. It was a good thing we didn't straddle it.

It's amazing how marriage, children and turning thirty matures a man.

Back at camp, we wrapped up the mission. It had only been two weeks, but we were ready to get back. I suspect if you would extend that to a year and throw in combat, that's what our reservists and guardsmen do. I was most impressed with these guys and gals. Handling family, deployments and their civilian jobs is quite a lift.

My immediate PSU boss CDR Robert Grabb realized I was newly married, and while there was a policy that everyone overnighted in the barracks at Fort Eustis during drill weekends, he let me sneak off as long as I was back by reveille.

After settling back into the Portsmouth job, I received an interesting call at work from headquarters.

"Can you go to Kazakhstan next week?"

Short notice trip to Kazakstan. Where the heck is Kazakhstan?

The call came from one of my old colleagues on the International Affairs staff. There was an emerging threat and opportunity to connect with a country that had been a previous United States adversary. With the recent collapse of the Soviet Union and the Chernobyl accident ten years earlier, there was a growing concern over nuclear materials in the former satellite states. Kazakhstan topped the list. While it's an Asian country, and I fancied myself as quite the world traveler at this point, I actually had to look on the globe to find it. It's the ninth largest country in the world by land mass and the largest landlocked nation, but its population was less than the city of New Delhi. Our destination was the capital city of Almaty, which is in the east, almost at the border of China and Mongolia. Again, why would we send Coast Guard to a landlocked country?

The southwestern corner of the country bordered the Caspian Sea. While nations claim a portion as a territorial sea, there is also a jointly managed area in the middle. Russia, Turkmenistan, Iran and Azerbaijan also shared a border with the Caspian Sea. Several were interested in getting their hands on nuclear material, and none were big fans of western democracy.

The Kazakhstan military had approached the US Coast Guard for advice on how their service could set up a legal architecture to permit their Guard Force to patrol and suppress smuggling on the Caspian. Our headquarters had identified a very sharp lawyer for the mission, but he was not excited to make a solo trip to the other side of the planet.

So, given the chance at more frequent flyer miles and a country I had not been to, I jumped at the opportunity. Let me not sugar coat this; the flight sucked. It was worse than the one to India. After two days of travelling, we eventually made it to Almaty. Our meetings went well, and I was, for the most part, a sidekick jumping in occasionally offering thoughts on patrol and boarding techniques and smuggling tactics that we had seen in neighboring countries.

The Caspian Sea can be easily exploited by smugglers, especially if government responses were not coordinated. Our host, pleased with the progress, took us on an escorted tour of the city. At that time, they were very much influenced by their neighbors Russia and China. The population was a mix of Asian and Caucasian. While I think they were genuinely happy to now be self-ruled, they were extremely cautious with big brother sharing the border to the north.

This was the farthest I have ever been from any ocean. While the lawyer carried the day, it was a good experience that caused me to brush up on proliferation security initiatives and weapons of mass destruction critical in future assignments. Plus, I got this nice letter from the big boss.

The best moment of my life came during this tour. I received the frantic call from my wife that her water had broken with our first child. I didn't have to say a thing; my boss, being a decorated pilot and a father himself, saw my face and told me to go. She was three weeks early, and I still had some final planning left to do. I had a flashback to the autobahn from my cadet days and made it to the house in record time.

Obviously, the ladies do all the heavy lifting, but, dads, there is no feeling quite like the very first steps that you take while leaving the hospital carrying your first newborn in your arms.

Lessons Learned:

Deploying guardsmen and reservists earn their paychecks.

Wearing issued fatigues is cool; so is shooting a grenade launcher.

Turkey is a NATO ally, yet it's in a precarious geographic and religious position.

Use care when choosing suntan lotion.

Learn how to take a power nap.

You hear strange and sometimes unpleasant noises when in a tent with thirty guys.

Our enlisted officers-in-charge are amazing.

Have a kit ready to deploy. Keep your passport current. Jump on any opportunity not involving a desk.

Know your role in each mission. Staff work is not sexy but necessary. When you're an operator, avoid bad-mouthing staffers! You will be one soon.

You can learn a lot about a person by seeing how well they do tasks they don't want to do.

Do your research on smuggled nuclear material. There are people in our government (Defense Threat Reduction Agency) that counter nuclear material proliferation for a career.

The Caspian Sea has several border countries that are not in the United States fan club.

Associations with the right people can be life changing. The opposite is true as well.

Become a go-to person. Have multiple ways to contribute. If your boss expresses interest in something, you should be fascinated by it.

There is no greater yet scarier moment in life than leaving a hospital with your first child.

Chapter 19

Group Commander During 9/11

THE COOL THING ABOUT making lieutenant commander is that *commander* is in the title, and you can now retire even if you fail to promote again. It's effectively tenure for a military officer.

It had been a big gamble pulling myself out of the running for positions on larger ships. That had been the normal career path, so several people said I would be committing career suicide if I made a switch mid-career. Based on three great ships and lots of action, I didn't think it was going to get much better. Two other factors weighed in. I saw what the stations and groups were doing in my current staff job, and instead of manning yet another—mind you, larger—ship, if I switched, I had a chance to manage multiple ships, boats and aircraft in a zone of responsibility. I thought that sounded cool; it would be new and challenging. Plus, Nye and I just had our first son, and I selfishly did not want to miss him growing up. We were also eager for another child.

After asking the assignment officer to pull my name from consideration of future ships, I made an even bolder move and requested to be screened for ashore command. The reply was rather quick. They said no way since I hadn't even served on a group staff. However, I reminded them that I had been on two patrol boats working for groups and been the manager for them for the last four years. I thought that might sway them, but it didn't. The detailer politely said I needed to submit a more realistic dream sheet, so I complied and asked for group executive or operations officer positions.

Time went by, and the detailer called me back out of the blue, telling me to resubmit and request group command. I'm not sure what changed as I had mentally moved on. I can only suspect someone of influence weighed in on my behalf. I think it might have been VADM Roger Rufe, the current Atlantic Area commander.

Regardless, I screened for ashore command. There were three group command jobs open for lieutenant commander in 1999: two on inland rivers and one coastal. I put in for all three but preferred Group Eastern Shore on the coast. Staying in the Mid-Atlantic kept us close to family and made sense as I had been the program manager for this group and knew many of their challenges and the surrounding commanders. In fact, my Academy classmate, Todd Watanabe, was the current CO. I must have made a convincing argument because I got the job in Chincoteague on the Delmarva Peninsula. They call it DELMARVA because it includes Delaware, Maryland and Virginia.

The main command resided in the small island town of Chincoteague, Virginia, known for its herd of wild ponies. It was an ideal rural spot for a young family. The Coast Guard was the biggest employer in the town of 3,000, and I think we were the third largest organization in Accomack County behind the Tyson chicken packing plant and NASA.

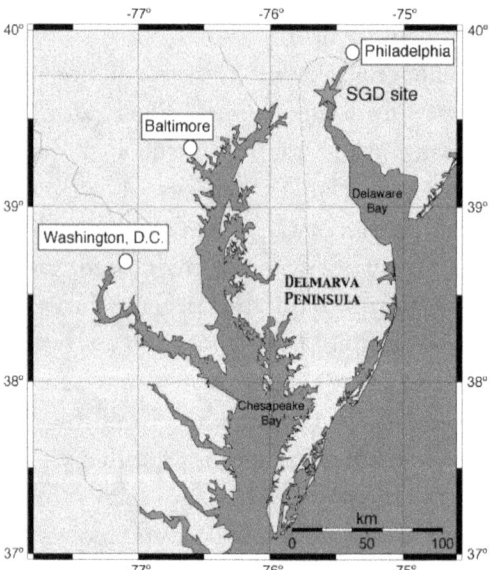

Crossing over the Chesapeake Bay bridge-tunnel to DELMARVA was like going back to the 1950s. Everything slowed down, people were friendly, and the family was more important than work. As crazy as it sounds, I felt the weight lifted off my shoulders as I crossed that bridge and paid the $14 toll. Maybe it was the wallet getting lighter, but that toll was high enough to deter hordes of tourists from visiting.

Group Eastern Shore Area of Responsibility covering ocean coasts of Delaware, Maryland and half of Virginia

Rumor had it that a flag officer had not visited the group office in seven years. That would change, but after a demanding job in Portsmouth, I looked forward to this calmer way of life. We had a bowling alley for recreation, and many areas for fishing, crabbing and clamming. If that wasn't enough, hunting was an option. I upgraded the Jon boat to a ski boat as I wanted to explore all the waterways and boat ramps.

When Nye and I went house hunting, it was slim pickings, but we landed a senior chief's house in Pocomoke, Maryland, just north of the Virginia line. It was pretty small, slightly less than our previous house with 1,100 square feet and one and a half baths. My son walked the trail to his elementary school, and we left our doors unlocked at night. It was only a twenty-minute drive over the state line and past Wallops Island to get to work.

Wallops Island was home to NASA and a Navy facility that included a joint housing area and a space launch pad. This wasn't like Canaveral. Most people never leave the main highway that runs north and south up the Eastern Shore. It seemed very out of place to have this huge runway and rows of enormous satellite disks on the left while driving down the road to the causeway that led to Chincoteague Island. Continuing over another bridge, the Assateague National Seashore waited. It was a protected beach area that extended well into Maryland, and that's where the wild ponies roamed.

Each year, the volunteer fire department, known as *Saltwater cowboys*, rounded up these ponies and drove them to make the swim to Chincoteague Island, where they were auctioned off. The proceeds went to funding the Volunteer Fire Department and also kept the herd size viable as there were no predators.

The biggest traffic day of the year in Chincoteague. Saltwater cowboys drive the wild pony herd for auction. (photo courtesy of DVIDS)

The command area of responsibility was a decent size and covered all coastal Delaware, Maryland and half of Virginia's ocean coast. The Assateague National Seashore covered about one-third of the middle area, so we worked closely with the Park Service. That gave me flashbacks to my high school days with the Youth Conservation Corps. The group had five units that had their own enlisted commands. It included four stations, an Aids to Navigation team, and even a lighthouse. Many men and women, especially the senior enlisted, were originally from the Eastern Shore and wanted to return for twilight tours. In all, there were four officers including me and a total of 250 people that included active duty, a civilian, reservists and auxiliarists.

Besides receiving my acceptance letter for the Academy, this was one of my first contacts with the Coast Guard Auxiliary, who were uniformed volunteers who had boats, radio stations and even airplanes. One pilot was a retired chief of the transit police from New York City. These folks proved to be great patriots and mentors for me. Our missions on the Eastern Shore were primarily maintaining the waterways and conducting search and rescue and law enforcement.

Before taking command, I had a challenge right off the bat. Then-CDR Dean Lee, my old carpool mate, was the group commander in North Carolina and drove up to see me take over. You see, he requires his buddies to work in a word of "his choosing" into their change of command speeches. Without letting the presiding official know what the word was, it had to be carefully woven into the remarks while staring at one of our buddies in the audience. I don't think anyone in the audience paid any attention to my remarks, but several were there just to make sure I completed my mission. For me, he had chosen the word *addendum*.

Fortunately, we had ruled out references to body parts. At the end of the incoming CO's remarks, there is a normal canned saying: "XO, take charge of the unit, all standing orders remain in effect." But I modified it to "all standing orders and addendums remain in effect." The presiding official on stage with me was VADM Rufe, the current Atlantic Area commander. I delivered the line looking at Dean Lee, my fellow group commander. I was told that Admiral Rufe, who was a combat-decorated veteran and patrol boat skipper during the Vietnam War, had a priceless WTF expression.

My first XO was none other than Quique Ortiz from the IMLET. A charismatic, good-looking guy (he asked that I put that part in) convinced me to let him sneak down to Colombia on a bit of an unsanctioned, interagency, drug mission. The action we were used to was dramatically slower in Chincoteague. I approved, and he quietly got the drug bust.

To be effective, you really need to try new things, similar to my time in Savannah. For example, most station small boats are limited to operating within ten miles of shore. The largest boats can go fifty miles out, but that is rare, and it would be out and back due to fuel and accommodations. Patrol boats are designed for offshore areas, despite often wanting to stay in a protected river or bay. Lawbreakers are aware of this limitation.

We got a waiver to tow one of the smaller station boats with a larger patrol boat going beyond the fifty-mile limit (in good weather). Then the crew of the small boat went just over the horizon to board unsuspecting fishing vessels.

The word got out to the fishing fleet that station boats had new operating parameters. Deterrent achieved.

We also started taking the station mascot out on patrol. Boaters with drugs didn't know the dog's capabilities, and it was enough to spook them into dumping their stash.

Finally, we received intelligence from the local police that there was likely a methamphetamine lab in the area. Because of the clandestine nature and smell of manufacturing this illegal drug, we recognized the lab would want to be in a remote location. The police had advised us that remote island houses or even large duck blinds could be the site of this lab operation. We called in Coast Guard special agents and specialized surveillance teams who were authorized to deal with the situation.

The engineering officer at Group Eastern Shore was Chief Warrant Officer Kerry Bowden. Chincoteague-born and bred, funny, and well-liked, he got things done and knew everyone. The operations officer was Chief Warrant Officer Dave Lewis. He was the father figure and basically ran the group. He, like Kerry, was a legend both in Chincoteague and the Coast Guard. His dad, known as Popsicle, was also a former Coastie and resident of Chincoteague. Dave, who had multiple commands under his belt by then, quickly took me under his wing. A saying he left me with early and I share today in my command courses is, "It's easy to be a visionary if you don't have to do the work."

The group had been all-in on the total quality management stuff. Todd, who was my predecessor, used to say that if you want something done, measure it. He was right, but it consumed countless long hours of staff paperwork work that was compiled and bound into fancy binders. Unfortunately, I'd been on the receiving end of that at the district. When we dropped from a staff of eight to two, we didn't have the time to go through it. We killed that off and made a few fans in the process.

That year, due to a spike in gas prices, the Coast Guard was very limited in the number of hours we could run our boats. In addition to our normal rescue and law enforcement missions, NASA often asked us to support their launches, ensuring their coastal launch areas were free of boats. Having to scrub a launch because of a fouled range conceivably costs millions of dollars in taxpayer money. With the district's permission, I met with NASA representatives and worked on a reimbursement plan that went into the unit coffers. That allowed us to cover all the NASA launches, obtain an additional boat and build reserve boarding teams. While that didn't seem like a big deal, we didn't have any scratched launches, and it would pay off during 9/11.

There were other challenges in the local government just before my arrival. The mayor, town manager, and chief of police were all relieved of their posts based on some perceived ethical issues when they traveled to Key West for a hurricane planning session. That left the city engineer effectively in charge of managing the hurricane response. He, in turn, reached out to us for help. Sure enough, we were immediately challenged with an approaching hurricane. Based on great staff and local connections, we assisted the town in manning their Red Cross shelters and a standby boat to ferry people off the island to the mainland hospital. In a small town, everyone steps up. The unit got a nice award for this mission.

Within Chincoteague's police department, I had a friend and confidant that recently retired as a warrant officer from the Fifth District and was now a police officer. He was a crusty old Chincoteague boatswain mate who had served at Group Eastern Shore in a previous tour. He was a valuable mentor for me at both the Fifth District and during this tour in Chincoteague. That said, we screwed with each other mercilessly as officemates in the Fifth District, so I thought nothing of continuing that relationship. I went into the police station from time to time when he wasn't there and used his ink pad to leave fingerprints on his desk with a note that read: Guess who?

In a small town, relationships are everything and having friends who were born and raised there made all the difference in the world.

Building teamwork, enhancing local environmental reputation, and getting dirty can be fun. There was a long causeway and one bridge to the Island of Chincoteague. Tourism is important, but those visitors also dump their trash on the way in. We adopted the causeway. We borrowed ten kayaks and took a morning to collect a dump truck load of debris every six months. This helped form a bond that paid dividends.

Things were going well, but not without challenges. As the commander, it was my personal duty to conduct next of kin notifications for those who perished at sea in my area of responsibility. That was VADM Rufe's rule, and while it was by far the toughest part of the job, it was brilliant. This meant the top person had to be intricately involved with the details of the case because they had to sit down with the families of the lost. This, in turn, meant the commander's staff and all subordinate commands were also intricately invested.

One of those next of kin notifications occurred at Indian River, Delaware. Two seamen were in the station's boat basin at midnight fishing from the pier. The station was perfectly positioned just inside the jetties to respond offshore. The seamen observed three civilians in their small, center-console boat trolling inside the inlet marked by a bridge. Breakers often form at the inlet as the tide funnels through the narrow jetties. Not having much luck, the civilians decided to turn on their white light to change out their rigs. Once re-rigged, the boat operator hit the throttle to reposition the boat for the best new drift. With his night vision impacted, the driver crashed into the creosote wall near the station. He was ejected, and the other two were incapacitated in the boat drifting toward the bridge and breakers at the mouth of the inlet.

The two Coasties sprang into action and affected the rescue of the two in the boat and got them to the hospital. Meanwhile, the rest of the station was now up and actively searching for the missing civilian who went overboard.

The station supervisor, through the command center, advised me that they found hair and suspected brain matter on the wall and that the two civilians in the hospital had called their families in West Virginia. Upon hearing the news, the families were immediately enroute in two vans to arrive at the station in two hours. I was there in an hour and a half. We were able to get them to go to the hospital first vice witness diver recovery operations.

In another case off Chincoteague, the young lady on the far left in the news clipping (Seaman Bening) was involved with the surf rescue of the motor yacht *Princess Leia*. After retiring, I used this example of risking lives to save lives and property in the Command Course. The officer in charge of Station Chincoteague at the time was Chief "Axe" Shellenberger who almost got me busted by MPs in Puerto Rico. Under his direction, and at great risk, the crew entered the surf, convinced the crew to abandon the ship just before the vessel broke up in the surf.

Daring rescue leads to medals for Coast Guard crew

Last Wednesday, six Coast Guardsmen at Station Chincoteague received medals for their "superior performance of duty" in a rescue operation off the coast of Assateague this spring.

Seaman Frances Bening, Fireman Walter Argüdo, Petty Officer First Class Bruce Birch, Petty Officer 3rd Class Matthew Carlisle and Petty Officer 2nd Class John Deeson each accepted the Coast Guard Achievement Medal presented by Craig Shellenberger, the officer in charge at the station.

Petty Officer 3rd Class Duane Buckner received the Coast Guard Commendation Medal for his participation in the rescue. Buckner was also named the Delmarva Lifesaver of the Year earlier in the fall.

Present for the ceremony were Lt. Cmdr. Mark Ogle and Lt. Michael Baroody, both of Coast Guard Group Eastern Shore.

The awards stem from the night of April 17 when a Florida couple became stranded aboard their stricken yacht in 10 foot breaking surf off the coast of Assateague.

With gale-force winds and fog preventing an air rescue, the crew maneuvered the station's 44-foot lifeboat through hidden shoals to reach the yacht which was in iminent danger of capsizing.

Because of the shallow water depth and relentless breaking surf, one false move could have sent both vessels and crews into the water where even trained swimmers would have struggled against the conditions.

Fortunately, the yacht's crew were rescued in the nick of time. Minutes later, the pleasure craft broke apart and its remains were found the next morning washed up on the rocks of Wallops Island.

The citations that accompanied the medals praised the Coast Guard crew's "exceptional fortitude, dedication and devotion to duty (which) are most heartily commended and are in keeping with the highest traditions of the United States Coast Guard."

Five Coast Guard Station Chincoteague members were awarded the Coast Guard Achievement Medal last Wednesday for their role in a rescue this spring. They are (l-r): Seaman Frances Bening, Fireman Walter Argü do, Petty Officer 3rd Class Bruce Birch, Petty Officer 3rd Class Matthew Carlisle and Petty Officer 2nd Class Joh Deeson. Petty Officer 3rd Class Duane Buckner (not pictured) was awarded the Coast Guard Commendatio Medal for his efforts in the same incident.

Courtesy of Chincoteague Beacon

In the Command Course, we just have captains and senior commanders, so I randomly called on one female commander to explain what she would do if placed in a similar circumstance.

CDR Frances Smith said, "That's easy. I'd do the same thing I did twenty years ago. The seaman on the left was me."

One of the finest officers in charge I ever served with was Master Chief Lou Fisher who ran the station at Ocean City, Maryland. He also retained oversight responsibility for the smaller unit at Indian River, Delaware. With thirty years in the Coast Guard, he had more than his fair share of heroics, but like all of us, we were getting older, and this is a young person's job. It was time for him to retire.

As an LCDR with half the years he had, I reached out to my new boss to see if he would preside over his retirement ceremony. I would, of course, preside over the change of command portion. Sure enough, the current Atlantic Area Commander VADM Thad Allen agreed. He was in an incredibly busy job, but he also knew how to prioritize work.

I knew Admiral Allen from my earlier staff tour. Despite being the Atlantic Area Commander, he also took his responsibilities as the Fifth District Commander very seriously.[1] He visited every one of Group Eastern Shore's units. I believe that was true for all the units in the district. It didn't hurt that his executive assistant was currently CDR Cari Thomas. I was the best man at her wedding. Isn't it cool how these connections work in a small service?

So, the big day was happening, and I helped write the admiral's speech. I had money down that I could make the tough guy—Master Chief Fisher with his gray flat-top haircut—get misty-eyed. He delivered! But in the middle of the ceremony, the ready boat crew sprinted down the basin with an Ocean City paramedic. It wasn't unusual as this was Ocean City, Maryland, the group's busiest station running over 400 search and rescue cases a year.

The station officer of the day came behind me on the stage and whispered in my ear that an offshore recreational diver had come up unresponsive. I turned to VADM Allen and asked if he would cover the reception so I could work search and rescue. He agreed. I was able to intercept the small boat when it was mooring up.

The diver was a middle-aged man, deceased in a body bag on the boat. The dejected crew had done everything they could to revive him. I looked at his hand and saw a wedding ring. The local news had overheard the radio traffic and had already picked up the story. The man had been on a dive boat out of Indian River. The operators and other passengers had no idea what his name was. He had apparently left his wallet in his car, and the only personal item they had was a set of car keys. I knew it was my responsibility to contact the next of kin, and people want to hear it from the authorities and not the news.

I got in my car with his keys and drove quickly up the coast to the parking lot where the dive trip had begun. The parking lot was full with a couple of hundred cars, not all for the dive boat. I remembered a trick I had seen on TV. If you hold the key fob to your chin, your head acts like an antenna and extends the range. Sure enough, I heard a car beeping. I went through the car and found his wallet along with what looked like a list of numbers for his wife. It had been several hours since the accident was covered in the news. I called the number, confirmed her name, and

---

[1] Admiral Allen became the National Incident Commander for Hurricane Katrina and the Deepwater Horizon Oil Spill and served as the Coast Guard Commandant.

said I had some very, very bad news. After I briefly told her, she became extremely upset. I guess her manager or co-worker grabbed the phone from her and cursed me out. That was the worse one I've ever done, and I've done over twenty.

A seaman from the station followed me in the man's car to his home in the DC area. There I was able to sit down with the family and provide everything I knew, including the husband's location. An investigation was ongoing, looking at the dive equipment and the vessel.

The wife and family were very kind during this follow-on discussion. From that point on, I have made it a point to become an expert in delivering this life-changing news. That said, I don't think I'll ever get used to it. I remember everyone I've ever done, and they're all different.

Next of kin notifications aren't the only challenges you face as the commander.

I received a call on a Sunday from a state senator in Delaware (not Joe Biden) asking if I was available to attend a little town meeting in the community center to discuss ongoing shoaling concerns in Delaware.

When a senator asks you to do something, the right answer is, "of course, yes," even though shoaling is really an Army Corps issue. My instincts were good on this one, and I decided to reach out and invite my buddy from the district, John Walters. He was probably ten to fifteen years my senior and ran the Aids to Navigation Program with Captain Cook for the district. He was a smart, retired Coast Guard officer who was an expert at his trade.

I have a rule of thumb that has always served me well. If you go to an event, especially one you may speak at, go early and get the lay of the land. I was early, and it was not what I expected. The Delaware town hall was packed with an angry mob. There were also at least three TV crews and a row of seats behind the microphone. One of those seats had my name on it. I quickly linked up with our buddies in the Delaware Natural Resource Police to game plan the session. For whatever reason, the Army Corps, who is normally on the ball, was not at this event. They were smart.

The meeting started when the senator took the microphone and said, "I would like to introduce the Coast Guard commander who has decided to pull the buoys and close your port down. LCDR Ogle, the microphone is yours."

As I walked up, I wondered what happened to that nice man who had called yesterday.

I think this was a pivotal moment for me. I said, "To start with, the reason we are here today is a storm has silted in our channel. This is no one's fault. As far as the buoys go, we have an aid to navigation boat with a small crane that can barely get in there when there is good water. If we didn't go in and pull those buoys when we did, you'd have sinkers and chains lying just below the surface. Then when grandma and the kids are returning from a nice day on the water, she could hit that hazard and send them through the windshield. I made the command decision to pull them for all *our* safety.

"In addition, we can get back to normal once the Army Corps comes in with dredge systems. Having the chain and sinkers on the bottom will greatly hamper and delay that effort. As for the port, we have not closed it. We have also teamed up with your very own Delaware Natural Resource Police to procure and jet in temporary PVC pipe to form day markers showing the best water until the Army Corps can get in and dredge. Any questions?"

No questions from the crowd but they offered a profuse thank you for everything we were doing.

I said, "Back to you, senator."

His reception was not as pleasant.

Our reputation on the shore was very important. When you're a commander, your job is to protect the Coast Guard brand. One station had a bit of friction, primarily due to fisheries enforcement cases. Bruce McMasters, the town manager in Wachapreague, was also a Coast Guard auxiliarist and a good friend. We teamed up to hold a festival with boats and a helicopter. A big deal for a small town. I was named the grand marshal. We set up a five-kilometer run for one hundred people.

I encouraged Nye to run it. She finished in last place. I still remember the piercing look she gave the grand marshal as she crossed the finish line.

This event, along with a couple of aggressive boarding officers rotating out, managed to improve relations slowly.

All these events pale in comparison to what happened on September 11, 2001. The day started very early. On the Eastern Shore, housing was very limited, so we had government-owned and managed housing. Our folks in Chincoteague lived in a World War II vintage neighborhood in Wallops Island on NASA property. They were built hastily in the 1940s and were expected to be temporary dwellings. Sixty years later, they were housing Coasties. The same goes for the age of some of our ships, by the way.

The housing units were in such bad condition, some were condemned. Our civilian housing officer, Mr. John Peppers, was a retired Army captain who managed both the unaccompanied personnel (our barracks) as well as these aging houses for familes. The guy worked wonders with duct tape, but it was a losing battle. The tenants were actually paid a stipend to live there because they were classified as substandard. We did what we could, had neighborhood parties and yard of the quarter competitions, but they still were in terrible condition. This became one of my biggest priorities that had been started with Todd Watanabe and his predecessor. But we were the little guy when it came to the district. Most COs were full Captains. We were somewhat out of the site and out of mind. Therefore, every report I sent to the district included something about the houses' deteriorating status. We often included pictures, and when that rare visitor came, we made sure the neighborhood was part of the tour. In addition to the primary government housing area for the Chincoteague-based Coast Guardsmen, we also had four government houses in Maryland supporting Station Ocean City.

The morning of September 11 started at 0500 when the command center advised me that there had been a government house fire in Ocean City.

My first question is always regarding people. Was everyone accounted for? Was everyone okay?

The petty officer who lived there was on duty at the station, leaving his wife, young son, and their baby alone. The young son was the first to sense the fire. He woke his mom and helped get the baby out unharmed. Thank God! We later recognized that young man for his heroics.

The fire department responded, but the damage was severe to a point that they recommended it be demolished. Since I was already up, I headed in early to work and made several calls alerting the district to the fire. A few hours later, I was in an all-hands training when the officer of the day rushed in. It was perhaps 0900.

"Captain, you got to see this."

We watched the planes flying into the towers. The clip we have all seen repeated over and over again. Then the plane hit the Pentagon. At that point, I think everyone knew we were under attack.

I was in command of the Coast Guard in portions of three states, and we were under attack in the homeland!

My granddad recounted his time in Pearl Harbor, but never in my wildest dreams could I have fathomed this happening. I called my wife, who was eight and a half months pregnant with our daughter Storm, and then I called for an all-hands muster. I wanted to find out who might have family members in the towers or in the Pentagon and prep them for leave immediately to care for their families. With great staff, they were well ahead of me as usual, and we started to set up barricades and posted security on the base, which was frankly wide open to the public. We had no gates at the front. I directed that all units take efforts to secure their facilities, then get out in their boats or vehicles with armed teams and show a calming presence. I also contacted our reservists and auxiliarists and put them on a short string; they would be needed. I was particularly lucky to have Jamie Bradshaw as our senior reserve officer. He was the right guy in the right place at the right time. His day job was as a Maryland Marine Police captain.

Fortunately, we didn't have any members directly at the points of impact. The effort to secure that spare boat and train up reserve boarding teams paid off, and we immediately deployed teams to New York City and DC.

My classmate Mike McAllister, who had relieved me on the *Key Largo,* was the operations officer for New York. His CO was in South Carolina, and the deputy was in the hospital for a medical issue. Mike was effectively the acting commander in New York during 9/11. They led an evacuation of over 500,000 people from lower Manhattan, even more than those evacuated during Dunkirk in World War II. Mike personally briefed President Bush.

One of the key leaders on the waterfront in New York City was a junior officer named Mike Day. Based on his actions to organize numerous boats and ferries to evacuate lower Manhattan, he became a bit of a folk hero. In fact, Tom Hanks narrated a video with him, and there was talk of a movie where George Clooney would play him (it never happened). Later, as a strike team commanding officer, he was my number two during the satellite shootdown case. Years later, he made captain and returned to Command Sector New York. As I'm writing this, he was just selected for admiral.

My post-retirement "work husband" was retired Captain Larry Brooks. On 9/11, he was the captain of the port in Hampton Roads, which was getting a lot of attention as it was the home to the largest naval base in the world. His deputy's brother was one of the victims in the twin towers. Larry immediately sent him off to comfort the family. His number two stepped up, but shortly after that, he had appendicitis and went down. Then number three stepped up, Pete Zohorsky, and they carried the day. Pete and I later became shipmates.

This was typical of all Americans. They all wanted to do something. But we were still trying to figure out what that was. The district worked miracles to get the Ocean City family displaced by the house fire put up in a hotel. Then the reservists and group engineers came to me and said they wanted to rebuild the burned house. This was unprecedented, but I told them to do it. Over the next couple of weeks, they did a phenomenal job.

I think, in retrospect, rebuilding something torn down on 9/11 was more important than just a house. On my departure, they presented this string-art masterpiece; the frame was from that house and is one of my most cherished possessions.

Frame was from the burned house in Ocean City.

There was no game plan for 9/11. We were on the eastern shore and expected more attacks. We were looking at the nuclear power plant in Delaware, the prevalent crop dusters that could fly over crowded beaches. We just did not know when the next attack would occur and what it would look like. But we expected it!

That first day was crazy, and admittedly, I did not sleep well. About thirty hours later, at 0400 in the morning, my wife's water broke—two weeks early. I should have been more ready after our son came three weeks early. The leave I had planned was obviously canceled. My new XO, Mike Baroody, was great and, having kids of his own, stepped immediately into the command chair. Nye and I dropped our four-year-old with friends and drove forty-five minutes to the closest hospital in Salisbury, Maryland. That's rural living for you.

We almost didn't make it. My wife's OBGYN doctor was stranded in Europe, as was my mom, who had made a rare overseas trip with a friend. All the flights had been grounded. My brother Johnny's wife Heather Ross was supposed to be on Flight 93, which went down in Pennsylvania on 9/11. By luck, Dr. Ross overslept that day and missed the flight. They even made a TV special about her.

When we got to the delivery room, there was no time for the epidural; the appropriately named Storm was coming.

My wife's response to the lack of pain relief was less than enthusiastic. "Great."

I'd seen that look before, at the end of the 5K race.

It was a pick-up game in the delivery room with a different doctor and Nye's family enroute from Newport News. I helped with the delivery! Storm was tiny but beautiful. I just wondered what kind of world she would now have to live in.

All the TVs in the hospital showed the planes crashing into the buildings over and over again, and then President Bush joined the fire chief on the rubble pile with the megaphone and gave a great speech. This was followed by the dump trucks racing to the rescue, flying huge American flags.

It was time to go get the bad guys!

Later that week, we had a special visitor wanting to go up to the Assateague lighthouse. We didn't normally do this, and only on rare occasions might we open on the weekend for tours. But this was Attorney General John Ashcroft. He wanted to clear his head and go up for inspiration. We did this quietly, and I never met him but could only imagine the pressure he was under after the 9/11 attacks.

During a stakeout in the marsh, these wild
ponies can give you a scare.

Lessons Learned:

Next of kin notifications are the hardest things a commander does.

Our nation rallied big time when attacked.

I like it when there's no playbook.

If policing a big beach or rafting event, have an observer in and amongst the participants to gauge reaction when the Coast Guard rolls in. The observers may see a drug stash thrown over the side or detect a possible terrorist with explosives.

If you think you have a meth lab nearby, don't go barging in—these things blow up. Call in the experts. If they take up a station in a marshy observation area, make sure they know about the deer flies and, more importantly, the wild horses.

Energy and resources are finite. Focus on the most important achievable goals.

Invite your boss to leave the desk and visit the troops. It's good for everyone.

If invited by a politician or media to come to a town meeting, be early and be prepared.

Bigger staffs actually create more work. Stay away from top-heavy organizations.

Chapter 20

A Chilly Swim Call

KEEPING CHINCOTEAGUE'S CAUSEWAY PRISTINE not only makes the island look more welcoming, but it also protects the critters, and if you find yourself swimming, it's nice not to see a full floating pamper. I have seen a few of those.

A month went by, and we were still trying to figure out what to do following the 9/11 attacks.

We had a couple of Coast Guard Auxiliarists who had been former CIA officers, and they helped identify threats in our area of responsibility. We were not a prime target like New York or Washington for sure. My area was mostly rural farmlands and beaches.

Because the 9/11 terrorists had pilot training along with the anthrax attack on the capital, we focused our attention initially on crop dusters that could fly down the beach. We also paid attention to the potential use of small boats as suicide bombs since they had used one on the USS *Cole* in 2000 and the tank ship *Limburgh* in 2002, both taking place in Yemini waters. We also paid close attention to the nuclear power plant in Delaware. Facilities were in the process of hardening their security. The post office started radiating mail. The airstrips, marinas and dive shops put up fences and cameras. While we were probably not the target of the attack, we could certainly be a good launch point.

I knew I should provide a commander's intent, so I created a point system for the units that I felt would effectively focus their effort to maximize detection and deterrence. I wanted a system that didn't tell the subordinate commanders what to do or how to do it, but rather the outcome I wanted. This was much the same latitude afforded me by the district. We still had our normal missions to do, but there had been a substantial swing toward law enforcement and security.

I remembered back to the Operation Spotlight presentation at the Academy when the number one patrol boat captain had to sift through tons of boats looking for smugglers. He didn't want to waste time boarding and annoying legitimate folks. So, the point system went like this: each unit would need to get 1,000 points in a year. A contact—meaning come alongside and talk to a boat—was worth one-fourth a point. Boarding was one point. A more complicated boarding of a commercial vessel was three points. Any law enforcement cases like catching a fugitive, boating while intoxicated, or a fishery bust netted the unit ten points. Since I had a thing for drug smugglers, one bust was fifty points. Finally, catching a terrorist was one hundred points.

It became a game to see what station could accumulate the most points. I guess you could say it was like a quota system with flexibility. The units in the south, which were very rural with less boating traffic, mostly focused on fisheries. In the north, we caught drunk boaters and some personal quantity drugs.

One morning, a southern unit was working on an illegal fisheries case. They roughly knew where the boat would be, so they took their station truck, which had blue lights mounted on it, to see if they could spot them from shore first. They went down a dirt road, and there was a truck next to the beach with smoke coming out of it. The occupants saw the Coast Guard truck and threw evidence out the window; they were detained by police. The unit then observed the illegal fishing operation. They vectored in their small boat for the boarding. They earned sixty-three points that night!

With terrorists, we tried to anticipate asymmetric threats. We visited boat rentals companies, dive shops and airstrips and made sure we had a great contact network. But we needed as many eyes on the water as possible, so we procured floatable, foam, key chains with safety equipment carriage requirements printed on one side and a hotline number to our command center for suspicious activity on the other. These were passed out to all boats encountered.

About a month had passed since 9/11, and I was sitting in my office in Chincoteague when the search and rescue alarm for the co-located station went off. Out the window, I saw the boat crew hustling for the boat. While Chincoteague was less busy than Ocean City, collectively, we were still doing around 1,000 search and rescue cases every year. Station Chincoteague had about 200 of those. I always get excited when there is a case. A new ensign hovered in my doorway. We were officially allowed a total of four officers for the group. Ensign Etienne De La Riva was an over billet, and he was filled with excitement. He was as gung-ho or more than I was when I was an ensign, and that's saying something.

Generally, Eastern Shore gets pretty quiet in the fall and winter, with the exception of this fall, of course. It was October now; things were slowing down a bit. It was cooling off, and the leaves were changing. Our newborn daughter Storm was up most of the night, so I was probably slowing down too. I looked up and saw the ensign wanted to go on the scene. The word was a car had sped out of control at the Curtis Merritt marina just down the road and launched into the boat basin. They couldn't see the car from the surface. I heard the volunteer fire truck filled with saltwater cowboys and their ambulance taking off from across the street as well.

The only experience I had with submerged vehicles was the Coast Guard helicopter dunker training in Mobile, Alabama. As cadets, it was a summer, two-week program. Part of that training required us to be strapped into the simulated fuselage and flipped upside down in a swimming pool while effectively blindfolded. We learned to get a reference point, wait for the violent movement to end, then egress.

The good news in this case was the car was no longer moving, and it was daylight.

Seeing the anxiousness in Etienne's face and knowing I had passed on an unwritten rule to go on the scene (if you're not a distraction), I agreed to hop in the car with him and Dave Lewis.

When we arrived, Station Chincoteague's small boat was in the middle of the boat basin, and Petty Officer Bruce Burch was in the water. There were probably twenty to thirty people who had gathered at the edge of the marina to observe the unfolding spectacle. The fire truck had beat us on the scene and was unloading as were the police. Bruce, from the *Princess Leia* rescue, was now treading water in the middle of the marina and saw us. He called out that the car was below him. You couldn't see it as it was about ten feet down in the murky brown water.

As the region's search and rescue mission coordinator, I knew we were equidistant from Air Station Elizabeth City, North Carolina, and Air Station Atlantic City, New Jersey, so a professional survival swimmer in the aircrew would not arrive in time. Witnesses thought there were kids in the car. We were hoping for an air pocket, because it had been at least twenty minutes at this point. The water was also getting cold in October, which sometimes can extend the chances of reviving someone.

The fire department to my left had made it down to the dock and tried to outfit a person with their regular oxygen tank used to fight a house fire. That looked cumbersome and didn't look like it was going to work. I thought back to my lifeguard merit badge in Boy Scouts and that stupid cinder block drill.

They also gave us swim training at the Academy. While never a great swimmer, I had one thing going for me. I could hold my breath for two minutes. About that time, Dave Lewis, a seasoned warrant officer of almost thirty years, came over as I was taking off my dress uniform shirt. He gave me one of those looks like, *Boss, what are you doing?*

To make Eagle Scout, you must first earn lifesaving merit badge which included swimming down ten feet in a lake to bring up a cinder block.

I answered his nonverbal question. "This is exactly why I joined the Coast Guard."

I signaled to Bruce, jumped in and swam almost to his position. Then I did a jackknife dive over where he thought the car was.

Sure enough, Bruce was exactly where he needed to be. I could only see maybe two feet in front of me, but I saw the driver's window. It was bad news; it was rolled down, so the air pocket was unlikely. In the first *Jaws* movie, Richard Dreyfuss swam down to the boat that had been attacked. He saw a hole in the boat where the shark bit through the hull, then the head pops out. Well, that was what I saw just inside the driver's window, which admittedly startled me. The driver looked to be dead. I pulled out his left arm, but he was stuck. The seat belt was still on.

The car landed on the bottom almost perfectly, tires down. I was able to open the rear door behind the driver, swim in, reach over and release the seat belt. When I was doing this, I thought I saw some white objects in the back seat, but I was focused on the driver. I exited the vehicle and, this time, was able to pull him out through the window to the surface.

He was a big guy, but unfortunately, lifeless. I swam and passed him off to the firemen and then returned to the car and told Bruce I saw two white objects in the back seat. We both dived, praying it was not children. We got in, and, fortunately, it was just pillows.

The car was clear. The paramedics did everything they could to revive the eighty-year-old man but without luck. When they pulled the car out, the accelerator was frozen at over ninety miles per hour. Apparently, the gentleman liked to watch the boats come in and out of the marina and possibly had a medical issue.

I went home with a wet uniform, and Nye asked what had happened. I told her it was a busy day at the office. The group command center received a call from the district asking who went into the submerged car. They hesitated, thinking we might be in trouble for entering the car, but we were not. It was right after 9/11, and everyone was helping others because it was just the right thing to do. They sent a nice message. PO Burch and I made it into the paper.

Admiral Allen received his orders transferring him from Atlantic Area to headquarters as the Coast Guard's chief of staff. He encouraged then-Rear Admiral Harvey Johnson, who was the one star in charge of Operations Capability and Policy, to consider me as his executive assistant. I was off to DC and into the big league.

On a final note, years later, I was in headquarters with Admiral Allen, the Commandant, receiving an award on behalf of the deployable specialized forces from the Transportation Security Administration (TSA) administrator for a different mission. Admiral Allen referred to me as a "water retriever," recalling this case. I think he even enhanced the story a bit!

Lessons Learned:

If you have a fear of the water, deal with it as a kid. Practice holding your breath.

If in a submerged aircraft or car, wait until the violent motion ends, have reference points, and egress.

Take off your dress uniform before swimming in the ocean.

If you can't pull him out the window, then swim in the back seat and release the seat belt.

When a junior member is excited and wants you to go on scene with him, do it. Emails can wait.

Remember why you joined the service.

"It's easy to be a visionary if you don't have to do the work."
– CWO Dave Lewis

If your organizational title says "commander," it means you have commanding officers and officers in charge below you. They've earned the right to command. Tell them your desired outcome, not how to suck eggs. Have their back, and they will have yours.

Chapter 21

Executive Assistant to the Admiral

THE COOL THING ABOUT making the rank of commander is that you go from a plain bill on your uniform hat to adorning it with a gold design that looks like scrambled eggs. This gold indicates you have officially moved into the senior officer ranks by becoming an O-5. It's about half the number of scramble eggs as an admiral, but it's still scrambled eggs!

If I had to describe a job I would be really bad at, it would be one of those officers wearing the rope and running after an admiral. One of my mentors at headquarters, Captain Tom Ostebo, had a saying: "Only a dope wears a rope."

Well, he was wearing a rope too, and he was no dope. He became an admiral.

Running the phones, correspondence, and the admiral's schedule all sounded horrific to me. I'm not a big fan of following detailed policy, reading, or sitting in front of a computer terminal all day. But I had been given great jobs throughout my career, and I had just made commander, so I didn't have much of a choice. I found out that the one-star admiral I worked for had told the detailers he needed a full commander with both afloat and ashore command experience. For assignment year 2002, only four candidates met his criteria. The other three knew Rear Admiral Lower Half (RDML) Harvey Johnson's demanding work reputation. I didn't know him, nor did he know me. While the service was small, he was a pilot and had been assigned in different regions than I had. Harvey Johnson was a brilliant guy and the hardest working person I've met— perhaps on the planet.

Before officially reporting in, I set up a meeting with RDML Johnson. He carved ten minutes out of his seventeen-hour workday to sit down with me. That wasn't because he thought I wasn't that important; it's because he had the other sixteen hours and fifty minutes booked. I went in and almost immediately came clean, stating that I may not be the right person for the job.

He looked at me over the top of his reading glasses. "Do you want the job?"

The answer to that question when posed by an admiral was, "Absolutely, I'll do my best, sir, and I hope you can keep up."

(Just kidding on the second part.)

When I reported my first day, one of the first people I saw walking into the building was VADM Thad Allen who was now serving as the chief of staff for the entire Coast Guard. Having just been my former boss, he stopped me. "You know, I'm the second hardest person to work for in this building, Opie[1]. You are going to work for the hardest guy."

I thought, *Wow, terrific. My dreams are coming true!*

I knew he had something to do with the assignment.

We were in the Operational Capabilities Directorate, which made us responsible for ships, boats, aircraft, etc. Seven different captains ran programs and reported to RDML Johnson. Several of these captains pulled me into their offices immediately and offered their couches if I ever needed a place to lay down, rest or pray. One captain showed me his office. Instead of the traditional ball caps that lined the upper walls of a senior captain's office, he had empty Pepto Bismol bottles. Life in the headquarters after the 9/11 attacks was incredibly challenging.

As a one-star in the headquarters, RDML Johnson did not warrant a deputy or an aide. The executive assistant had to do a combination of those tasks. As a boot commander and first time in headquarters, I was in a delicate position working between an admiral and seven captains. I didn't really produce anything; I was more like a delivery boy, often with bad news.

The whole country was transitioning into the new normal following 9/11. This meant the traditional ways of doing business was over. That included standing up numerous counterterrorism units, buying new and faster platforms and getting into the national intelligence community. There was talk of standing up a whole new department like the

---

[1] He called me Opie as a reference to *The Andy Griffith Show.*.

Department of Defense for the homeland. I was slowly figuring out my way around headquarters when I got the word that RDML Johnson was swapping seats with the two-star, Rear Admiral (RADM) James Olsen, who was his counterpart in Operations Policy.

The White House made a formal decision to stand up the Department of Homeland Security (DHS), and RDML Johnson was to help lead this interagency effort and move the Coast Guard to the new department. No work in that!

RADM Olsen was also a pilot and a bit more laid back, and I thought that his Executive Assistant Matt Gimple would remain with the policy folks, and I would remain with capabilities. Nope. The admirals and their EAs shifted together. By this time, I was in sync with RDML Johnson, but I was not thrilled to craft and manage emerging policy.

RDML Johnson could be quite funny at times, but most people kept a wide berth because their plates were already overflowing with work. Plus, a couple of more admiral-selectees were brought in to help in our little slice of heaven. RDMLs John Crowley and Rob Parker were part of that team. Things certainly were busy. RDML Johnson was in before five a.m. and often left around eight p.m. I had to print perhaps 100 emails for him to process when he got home. I don't think he slept. He worked every Sunday but didn't expect me to work as he knew I had a young family. I did a calculation once, unbeknownst to him, of how much he made an hour. It was roughly the same as a civilian administrative secretary. He was not doing it for the money.

While I was a very junior executive assistant or EA, the other EAs, including Captain Tom Ostebo, were up-and-comers. The majority were captains. Tom was the EA for Operations. The EA for the Commandant was Captain Dave Pekoske.[2] EA for VADM Allen was Jody Breckenridge, who eventually made vice admiral. Roy Nash was the EA for Marine Safety and became a RADM and commanded the Eighth District. As EAs went, I'd be the underachiever.

I guess the EA job was one you would request if you're seriously interested in becoming an admiral. That job convinced me I was not. To be clear, I wasn't in the same league as those other EAs, but many in headquarters suspected I had requested the job. If you watch the TV show Simpsons, Mr. Bums's executive assistant is Smithers. I got that nickname along with the unflattering "Little Johnson" in the first few

---

[2] Dave Pekoske went on to be Vice Commandant and then TSA administrator.

weeks. That gave way to Lucifer or the dark angel as I delivered work on a Friday afternoon due first thing Monday. But that was in jest; they certainly did not want my job. I went to their office late in the afternoon, and we'd play the game.

"I know you're hiding in here . . ."

To give you a sense of a workday, it averaged around sixteen hours, but it was the pace and the pressure that really wore me out. After chasing after RDML Johnson, who walks faster than most run, we took a short detour to the head (bathroom). After taking dictation at the urinal, I locked eyes with a captain passing us. I told the captain in jest I officially reached a career low point. He laughed.

I only served with RDML Johnson the first year (seven in dog years). He took great care of me, and for that, I was thankful. At the end of the first year, he got his second star and received the coveted Seventh District Command in Florida.

His relief was RADM Jeff Hathaway, a ship guy like I had been. He, too, was very bright, a bit friendlier and more laid back. In his previous tour, he had been assigned to the Pentagon. In fact, during 9/11, his office and staff were where the plane impacted, but as fate would have it, he had a meeting that morning in the Navy Yard. Losing many of his staff impacted him deeply. He was a warm soul and a good mentor. He testified in front of Congress and frequently briefed the Commandant. Even when that didn't go well, he always had a joke and a smile on his face.

That year seemed to go by quickly, and I'm sure RADM Hathaway helped a bit on my next assignment. Probably a good thing.

I never really told this story, but as an EA in Washington DC, we used a government-wide motor pool service to carry admirals around the city. I had used them multiple times for RDML Johnson when he briefed President Bush on DHS stand-up updates. One day when RADM Hathaway had to brief Congress, for whatever reason, the motor pool car didn't arrive; we went into full panic mode. He didn't have his own car as he took the metro, so that left my ten-year-old, two-door, gold Camry. In his SDBs, which is our formal dress blue uniform, we sped off and made it in time. The driver service picked him up and brought him back that day. Upon returning to his office, he quickly recapped how *fun* it was to testify; he hung his jacket on the door and went to the head. That's when I noticed it.

It was covered in yellow hair. My yellow Labrador retriever Josy loved to sit shotgun when we were going out to the park. I jumped on the tape loops and did a quick clean-up job. No doubt RADM Hathaway would have laughed it off.

New housing for Group Eastern Shore had finally been approved and built. VADM Allen, who still served as the Coast Guard's Chief of Staff and handled all Coast Guard funding remembered that had been my top priority while in command there, so he and I flew in the Commandant's jet to the ribbon-cutting ceremony. He commented that we had to wear ankle bracelets when away from DC. He took the opportunity to ask me to do some pet projects, including helping upgrade the national command center as well as serve on a committee evaluating DHS regions. The head of that committee was RDML (Select) Rob Parker, who would eventually make three stars and preside over my retirement.

One of the few times I was permitted to leave Coast Guard Headquarters was when I was sent to Padre Island, Texas, to evaluate tactics being used against an increasingly aggressive group of cartels just over the border from Brownsville. We took an unmarked vehicle trip along the beach, and within minutes, we were made. This gave me some insight on cartels and new ideas for operating more effectively on our southern border.

Despite being in the big league and knowing many admirals, I was going to have great difficulty making captain without a master's degree, so I asked and gained permission to attend Naval War College at night. My first professor was Dr. James Carafano, a retired Army colonel and frequent Fox News contributor from the Heritage Foundation.

He really liked the Coast Guard for some reason, which made me feel at ease with the rest of the DoD students. That said, we had nearly ninety textbooks for the strategy class. The books literally filled the trunk of a Camry. RDML Hathaway said it's only a lot of reading if you do it. Ironically, many years later, while I was teaching a course in Thailand, the Thai admiral presiding over the graduation mentioned he attended the US Naval War College. He also said the exact same thing about the reading.

But Carafano's style was to come in, sit in a circle, and discuss the reading. I thoroughly read a small portion, perhaps one-half of one of the ninety books. Then when the time was right, I struck and recapped everything I could remember. Then went back to my dormant condition.

I called that filibustering, and it was a tactic used later to bail me out of a tight spot on a morning talk show.

For that class, I also had to write a thirty-plus page paper which was titled "Reenergizing the War on Cocaine." Luckily, as an EA, I escorted the DEA administrator, Customs and Border Protection (CBP) commissioner, and even Barry McCaffery, a retired army general and the current drug czar. So I had some great quotes from those elevator rides. Years later, someone I didn't know called me from an obscure intelligence agency asking me about something I had written in my paper. I had no idea the papers were shared!

I sent a note to Dr. Carafano when the unit I commanded broke 100 tons of cocaine in two years, referencing the paper! I got an "A," by the way. I think this should be the model for education. If you have to study, make it something you actually can use in life and work.

When receiving the orders as a headquarters EA, I thought it was a career death sentence. After those two maturing years, I now saw it for what it was: a much broader understanding of how the organization and the government really worked. I got major sympathy points and some name recognition which helps with future promotion boards. RADM Johnson spent only one year in Florida as a two-star and then became the three-star in charge of the Pacific. RADM Hathaway got a dream flag job at the Joint Interagency Task Force in Key West, which is the DOD, interagency and international organization tasked with fighting the drug war. My dream job was to be a commanding officer working for both these positions.

I had always wanted to be on a Tactical Law Enforcement Team (TACLET) after hearing stories from my carpool mates Keith Smith and Ed Guy, but I thought I missed my time. Lieutenant commanders led these teams. But that year, they upgraded the position to a full commander. The current commanding officer of the Pacific TACLET (PACTACLET) was Pat Dequattro. He had been one of the freshmen I put through boot camp, and we were in the same company together. He was a lieutenant commander and a rising star; he had even carried the football (with the nuclear launch codes) for the president.

To get the TACLET commander job, I had three big things going for me. My former carpool mate Keith Smith was now the commanding officer of the Maritime Safety and Security Team (MSST) in Los Angeles. When asked by the Pacific Area staff, he vouched for me. Then the two admirals, Johnson and Hathaway took their posts. I think that sealed the deal for me. I'd landed my dream job in San Diego.

As I was getting older and coming off the demanding staff tour, I wanted to get back into shape. Pat was a marathon runner who could bench over 300 pounds. These TACLET guys do two-hour workouts every day. For those assigned to a grinding job where you work like a dog and rarely get positive feedback, there is nothing like getting a command job in a beautiful place like San Diego.

Before departing, I linked up with Pat Dequattro and John Daly[3], the future commanding officer of the other TACLET in Miami. We had breakfast in the White House. Both Pat and now John carried the "football" with the nuclear codes for the president. That was pretty cool as we were sitting next to White House Deputy Chief of Staff Karl Rove. The White House Chief of Staff Andy Card made it a point to come over and shake our hands and thank us for our service. It's pretty cool to tell your kids where you had breakfast.

I survived DC! It was a happy moment to see headquarters in the rear-view mirror.

---

[3] I went to John's retirement in the White House, which United States Homeland Security Advisor John Brennan presided over. Daly also got a one-on-one farewell with President Obama and Michelle.

Lessons Learned:

When an Admiral asks do you *want* the job as his assistant, suspend being trustworthy for a moment and answer yes.

In war college, it's only a lot of reading if you do it. I do remember the best military leaders in history tended to be experts at the three levels of warfare: tactical (winning battles), operational (winning campaigns), and strategic (should we be in this war?). Many senior leaders struggle to let go of the tactical, leaving no one focused on the other levels. Know and fill your role.

When sending your admiral off to testify before Congress, make sure you get the dog hair out of the passenger seat of your car.

Moving rapidly through the hallways and offices of headquarters with a clipboard or red folder, even if it's empty, can give you a nice break.

If you eat breakfast in the White House, you also get souvenir M&M's.

It's great to see an initiative you championed reach fruition.

When supporting the author of the US Maritime Security Strategy following 9/11, look around the world, and see what nation has the best program. The Israelis are surrounded by countries that want to destroy them. By necessity, they are exceptionally good, and I had proof from my Colombia trip. We went to the Israeli Embassy looking for their security lessons learned.

Tactic: If you're a unit on the border, you can bet that your vessel movements and crew are monitored. Use that to your advantage and secretly bring in another force from the sea or a different location. If you can't afford an aircraft overflight at $5,000 an hour, consider renting a high-rise room overlooking a hot smuggling area for a week at $1,000. Make sure your legal office is onboard.

Just when you thought you promoted yourself out of a dream job, the rules might change.

Chapter 22

Third Command: Pacific Tactical Law Enforcement Team

IF I WERE TO describe a dream job in a great location, it would be warm, have beaches and mountains, relatively short but exciting workdays, and include time for workouts. It would have colleagues full of like-minded, passionate people. A plus would be free parking and not having to give speeches. Oh, and my boss would be a thousand miles away.

My dreams were fulfilled. By 2004, consolidation had occurred, and there were only two large TACLETs left. This was the reason the command position increased from lieutenant commander to a full commander. The PACTACLET was based in San Diego, and the second best—TACLET South—was based in Miami, Florida. While I like Miami, it's flat and hot and has mosquitos and hurricanes. Not to mention, a thrown rock can hit the admiral's office.

San Diego didn't have those environmental detractors and was over a seven-hour drive to my bosses in Alameda, California. This was a far cry from sitting at a desk outside the admiral's office in Washington, albeit it was the same admiral. VADM Harvey Johnson had gone from one-star to three-star in just over a year. He was now in charge of all the Coast Guard Operational Forces from the Rockies west to the Indo Pak Border, so even with his unrivaled energy and attention, we were rarely on his radar scope.

The only problem was my wife Nye had designed and built our dream house in Maryland, and we had only lived in it two years before getting these orders. I personally had seen very little of the house due to my demanding work schedule, but Nye liked it and the area. Her family was in Virginia just three hours away. They were less than thrilled about their only grandkids moving across the country.

When I knew it was coming, like all married military members, I started prepping the battlefield. I put out flyers on California like Disneyland, Hollywood and Sea World. I had my brothers casually make comments on how they longed to return to the Golden State. Then, when the time was right, I broke the news on the most miserable, sleet-covered DC day. My strategy wasn't perfect, and she wasn't initially enthusiastic, but she sensed my excitement and knew I'd been working really hard. Once we started planning to drive across the country, we all got excited. Moving as a single guy was simple—not so much when you have a boat, small kids, multiple cars, and a 100-pound, shedding Labrador. In retrospect, it was a good thing for the family. It was like spreading your wings, as my dad had said. I occasionally deployed while in San Diego, but nothing like I had during previous tours.

The PACTACLET was nestled near the end of the main runway for San Diego's main airport. It was considered one of the tenant commands aboard the Marine Corps Recruit Depot. As the home of the west coast Marine Corps boot camp, it was adorned by marching sections, obstacle courses, palm trees and yellow buildings with red roofs. Pretty much all cinder blocks. This wasn't an Air Force base. Nothing fancy, but it didn't have to be. It was southern California.

Marine Corps Recruit Depot in San Diego

My passion was counternarcotics, and I made that abundantly clear. Our mission was to raid the supply lines of criminal organizations poisoning America. Earlier in my career, I thought I was pretty good at the nuts and bolts of making drug busts. But in reality, at best, we were a nuisance seizing a few hundred pounds here or there. It was a drop in the bucket that the cartels factored in as lost inventory.

My role was to train and equip boarding teams for missions. I was not there to personally make drug busts or lead teams. That would be the LTJGs job, similar to my role on *Vashon*. This job required me to move up and operate at the operational level. This is when you string battles together into a campaign.

If you genuinely want to make a difference in the war on drugs, you have to get inside the OODA loop of your adversary. OODA stands for observing, orienting, deciding and acting, which is a war college phrase.[1] Essentially, the smugglers were inside our OODA loop. They were well-financed, organized and nimble compared to the US government, which, by design, is a slow bureaucracy. The smugglers were always two steps ahead of us.

But things were going on in the intelligence community, and tactics were developed which finally allowed us to jump ahead of our adversaries. If the United States could make a significant dent in the supply chain, and if the demand for drugs remains constant, then the street price goes up. If the price on the street goes up, logic dictates fewer people can afford to buy drugs.

When the ship doesn't want to be boarded, you may have to use a hook and a caving ladder. (photo courtesy of DVIDS)

---

[1] I completed the war college during this tour.

There was a hot debate on where the finite US counternarcotics resources should be placed: on supply or demand reduction. I told the troops it was above our pay grade. We'd let the academics wrestle with that. No need to waste energy in philosophical debates as our role was clear—reduce the supply.

Over the next three years, we seized 130 tons of cocaine with a conservative street value of $3.2 billion. Obviously, we factored ships in the equation, but it wasn't a bad return of PACTACLET's $4.5 million budget. And actually, one-third of that budget was dedicated to non-narcotics missions in the Middle East. The average cocaine seized by each of these young men and women deployers was 1.4 tons or 2,800 pounds of cocaine. We were breaking records.

While the counterdrug mission was job one, we also got involved with other law enforcement. That included taking down large, migrant-smuggling vessels in the Pacific.

Our boarding teams not only served aboard US Navy and allied ships, we also trained their boarding teams. This plaque was a nice gift from the Japanese.

Something I hadn't seen in the Caribbean was Chinese migrants. Generally, the smugglers had enforcers referred to as "snakeheads" mixed in the crowd who could become violent.

Our unit was also involved in boarding and arresting a high-profile, violent fugitive in the middle of the night off a cruise ship. If it was a dangerous mission, TACLET was on speed dial.

The TACLET organizational structure was unique and not well understood even within the Coast Guard. Of the roughly 100 personnel, it had an overarching staff of about fifteen. The XO was a lieutenant, and I was blessed to have first LT Rod Rojas and then LT Ed Songer. I also had a lieutenant as the operations officer, a talented training staff, and a few folks focused on administration and logistics. These folks were critical in shipping teams and real weapons globally. I knew all too well from my El Salvador trip that improper notification can lead to serious issues. Below the command staff was eight, ten-person, law enforcement detachments (LEDETS) lead by an officer in charge; either a lieutenant junior grade or a senior chief petty officer.

There were three levels of qualifications within the teams: boarding team member, boarding officer, and deployed team leaders. Rank wasn't as important as the ability to do the job. That meant even a petty officer could serve as the deployed team leader. It was never a problem within the teams, but it sometimes created issues with the host ship. These LEDETS normally deployed aboard US Navy or allied ships, either in the Eastern Pacific, Caribbean or the Middle East.

Despite the name Pacific or South, both TACLETs operated globally. We occasionally mixed teams or had personnel from our sister, deployable, specialized forces augment the LEDETS. Central Command (CENTCOM) had a Middle East theater requirement of two steady-state LEDETs. Generally, one had the primary responsibility for training Iraqi Marines, and the other conducted boardings or protected platforms in the Persian Gulf. The two traded off mid-deployment. As a general rule of thumb back then, LEDETs deployed for three months at a time.

As the TACLET CO, I met with the host ship captains, normally equivalent to my rank. I briefed them on the mission and tactics and conducted ramp-up training with their crews. It was not unlike the mission I had done more than a decade earlier preparing for Desert Storm. The only difference was that this was preparing teams for counterdrug missions—often a new experience for most of our Navy friends. They were used to deploying in battle groups, but when you're hunting smugglers, you hunt alone on the surface. They normally carried helicopters, and if lucky, they got fixed-wing air support. When our LEDET was onboard, they provided the boarding team, aerial observers and, eventually, aerial snipers. Aerial use of force was designed to provide warning shots first and then disabling fire into the outboards of non-compliant small boats.

In addition to sending these boarding teams to stage from US Navy ships, we also had teams aboard British, French, Dutch, Australian and Colombian vessels. While I didn't fancy myself a diplomat, I was no longer a novice when dealing with foreign militaries after the international training team, even though most of these countries were never visited.

You may wonder how the US Navy can enforce drug laws. Posse Comitatus prevents the Navy and the other armed forces from actively conducting law enforcement within the United States. That doesn't apply to the Coast Guard. So, when the Navy ship has a suspected drug vessel to interdict, the ship shifts tactical control to a Coast Guard admiral either in Miami or Alameda. The LEDET raises the Coast Guard ensign or flag, and our officer and petty officers affect searches, seizures and arrests. It's a neat concept and explains why our seizure numbers and quantities were so high.

Back in San Diego, I lived vicariously through the LEDET's success on the water.

Just as I had on the patrol boats, we very much competed with TACLET South busts as well as some friendly internal LEDET competitions to see who could seize the most. Beyond the drug busts, every three months, the Marine Corps hosted a Fitness Olympics and teams throughout the base competed. We usually won these events, because when we're at sea and not actively chasing a smuggler, we were working out. Every day in port, the two-hour mountain runs and ocean swims paid dividends. Our monthly PT test consisted of a 500-yard swim followed immediately by a mile and half run, then straight into push-ups, sit ups, pull ups and chin ups. Our ready for operations annual LEDET evaluation was epic, as it was designed by our training staff that included a petty officer who had twice been through the Navy Basic Underwater Demolition/SEAL (BUD/S) program.

While not always enjoyable, it ensured that if our folks tangled with an opposing force, they would prevail. It also ensured that our case packages were the best the district attorneys had ever seen. This training proved critical in the next few months.

Lessons Learned:

1. If you are in an arm-wrestling competition and lose, it might as well be to a Marine. Marines are great patriots and, I believe, have the best esprit de corps of any service.

2. A San Diego house overlooks the ocean and mountains, not requiring heat or air conditioning.

3. When they say as part of the TACLET fitness test, you must do five pull-ups and five chinups . . . that's with body armor and a gun belt. You don't fast rope in gym gear.

4. TACLET members work hard and play hard. Limit alcohol consumption at parties!

5. If a vice admiral apologizes publicly to your wife during the incoming change of command ceremony for spending too many long nights with her husband, be quick to clarify his meaning with the new troops.

6. The LTJGs are tactical (winning the battle = getting the drug busts). As a commander and TACLET CO, you're focused on the operational level (winning the campaign). Elevate to your role.

7. Tactic: Creating one interagency command center has many synergistic benefits; however, more people means greater opportunity for problems with operational security. Millions of dollars and physical threats can corrupt. If you're not getting busts on the border, consider conducting some special operations off the books but with top cover.

Tactic: If you don't have to maintain a ship, boat or aircraft, and all you do is boardings, you better be the best trained and equipped! If downrange, you're only doing a couple of boardings a month, so look for other opportunities. Customs was short-handed. PACTACLET augmented their eighty MPH interceptors (think *Miami Vice*) and made a bust almost every weekend. The unit also sent members to the land border checkpoints to improve search skills, to Lake Havasu for boating while intoxicated boardings (first year without a fatality), and our LEDET Officers in Charge to a special DEA operation to get insight into the bigger picture. Train your junior leaders!

Chapter 23

Largest Maritime Bust in History

MY PREDECESSORS HAD WITNESSED a growing level of threats and violence toward our boarding teams. Assigned to TACLET South in Miami, Nate Bruckenthal's combat death in the Persian Gulf was the most recent, occurring a couple of months before my assignment. There had also been gun grabs and a molotov cocktail thrown at a boarding team just prior to my assignment.

Petty Officer Nate Bruckenthal

The units experimented with new, safer tactics that were perhaps more effective. These included fast-roping from helicopters; use of covert caving ladders to climb the side of ships; flying drones; engaging precision sharpshooters from helicopters; and introducing a new concept of the unannounced nighttime boardings (UNB). This job was not for the faint of heart.

To this point, we had followed the Coast Guard boarding policy of making observations, then asking a series of standard questions, then do a horseshoe around the vessel, and direct the crew to do certain things before we stepped foot on the vessel. That alone gave them plenty of time to set up an ambush or booby traps, practice a cover story and destroy evidence.

The UNB tactic was controversial in most Coast Guard circles. It was straying far from our humanitarian roots. It was a very risky tactic yet doing what we always had done was becoming increasingly dangerous.

Only months after I arrived, we got an opportunity. Our sister

TACLET was on a Navy ship in the eastern Pacific and was covertly following what was believed to be a large fishing vessel with a big load of drugs. The fishing vessel *Lina Marie* was 1,000 miles west of the west coast of Central America. The smugglers probably felt making it that far from shore, they would likely evade interdiction forces. They were wrong.

I received a call from Rear Admiral Jody Breckenridge, who was now the Eleventh District Commander covering California South. We had been executive assistants together at headquarters. She was the tactical commander for the mission and asked if the TACLET South team could safely pull off this UNB.

I called Glenn Grahl, my counterpart Commanding Officer in Miami. Despite having fun with the competition, we made it a point to train the same way, and they were well-equipped to do the mission. So I passed that information back to RADM Breckenridge. But that's when things went sideways.

She was ready to use them, but TACLET South was an Atlantic Area unit, and their bosses at the time weighed in. They were uncomfortable with the plan. In fact, they stressed that UNB was not how the Coast Guard does boardings and recapped the policy of pre-boarding questions and having to illuminate the stripe.

Of course, I could have been a smart ass and said we're boarding from a Navy ship, so they don't have a stripe, but I let it ride. In the end, Glenn's TACLET South team was not allowed to conduct the boarding. I felt bad as stealing a bust from another unit is very poor form. It had happened to me once in Puerto Rico, and from that point on, I was committed not to do it. But you can't let the bad guys get away either.

TACLET South was well aware of the dynamics in play at Atlantic Area. I received a call directly from VADM Johnson, now the Pacific Area Commander and RADM's boss.

He wanted to know if PACTACLET could pull it off.

Senior Chief Todd Lafleur led the closest law enforcement detachment and had a great team. I spoke with Todd personally on the satellite phone, and both he and their host ship were confident. I called VADM Johnson and told him we were a go.

The boarding went well. There was a gun grab, but the weapon was retained, and the assailant subdued without injury. I used that example to explain why we work out for two hours every day.

USS Curts underway with our boarding team (photo courtesy of DVIDS)

Lafleur's team took over the boat rapidly somewhere around three a.m. Of course, we had permission in advance from the vessel's flag state. Bilateral agreements had been previously arranged to allow for quick prosecution of these cases.

Our team's intercept took time, so when this case went down, I was on the bow of the USS *Thunderbolt* in Iraqi waters. PACTACLET's XO at the time, Rod Rojas excitedly relayed that the LEDET had unloaded hundreds of bales of cocaine and were still pulling them off!

In total, it was fifteen tons of cocaine, the biggest maritime bust in history, and perhaps more importantly, they got the vessel's codebook without the smugglers getting a call off. With the codebook, we were able to locate and immediately jump the sister ship F/V *San Jose* which had approximately eleven tons of coke, which was later determined to be linked to the notorious drug kingpin El Chapo.

We seized four giant cocaine loads in a span ten days. Eleven of the forty tons were linked to El Chapo, a notorious Mexican Cartel leader.

The boarding officer testified in the El Chapo trial years later. That turned out to be the third-largest maritime bust in history. But we weren't done. Our guys got two more cases of eight and six tons, respectively. In a span of ten days PACTACLET seized forty tons of cocaine. Along with several other cases, it was a huge haul for the year. In fact, our small unit accounted for over half of all drugs seized by US law enforcement for that year.

We had gotten ahead briefly in the OODA loop, but that only lasted for a couple of months. During a boarding, one of our petty officers, who looked like a short version of the Rock, grabbed the rail, which sent him jolting back in the small boat. The smugglers had rigged a car battery to electrify the rails at night to repel all boarders. Then they started carrying dogs, not the loveable Labradors or Poodles. These were Pit Bulls and Rottweilers.

I want to be very clear that my role was in the rear with the gear. I was living vicariously through the LEDETs. When they returned from deployments, we watched their videos, which usually had great music overlaid. If they got a bust, I broke out the champagne as part of the brief. TACLET South, being less athletic, preferred cigars.

Based on our success, we received a lot of VIP visits, including congressional delegations. I even met "the Governator," Arnold Schwarzenegger. After I returned from the Middle East, I personally bought each member of the *Lina Marie* boarding team a nice plaque with a picture of the coke and a placard stating it was the largest maritime bust in history. VADM Johnson did me a solid and came down to present them. That's saying something, given his schedule.

The Marines were no-nonsense patriots and most gracious hosts to their two blue-suited tenants: PACTACLET and our sister unit aboard Marine Corps Recruit Depot, the Maritime Safety and Security Team. My sister unit's command team was great. Commander Mark Eyler was squared away and knew how to have fun. His XO was LT Bill Walsh. Bill had been a petty officer in Savannah and even set me up on a date with his sister. He was very trusting.

Every week, we attended the staff meeting with the Marines; they called it the "council of colonels." They enjoyed my current event drug bust stories. Upon his departure, we presented the commanding general a mounted set of handcuffs from the *Lina Marie* case. John Paxton earned his fourth star and served as the Assistant Commandant of the Marine Corps.

But not all of my tour was pleasant. The Pacific Command made me their go-to investigating officer. VADM Johnson's relief was VADM

Wulster. Late on a Saturday afternoon, I received a call that there had been a tragic mishap in Seattle. Petty Officer Ron Gill was assigned to the MSST out of Anchorage and was deployed to Seattle to conduct ferry escorts. Ron was the assigned bow gunner for the fixed M240 machine gun. He was ejected and run over during a power turn and killed by the boat's propeller.

Four Coast Guard boats in the area immediately responded but were unable to save him. My job was to fly to Alameda, California, meet with VADM Wulster, get an assigned attorney, then fly to Seattle and collect evidence, including interviews of everyone involved. Then I flew to Anchorage and met with the unit. This was extremely painful for everyone. I eventually briefed the parents and Ron's young pregnant wife with the Commandant, Admiral Papp, in Washington.

This was a vivid reminder that what Coast Guardsmen do is dangerous. Ron would not have been out there if the terrorists didn't attack us on 9/11. Nothing can be taken lightly, and everyone involved should have a say on whether to conduct a mission or not. Eventually, the MSST in Anchorage was decommissioned, and all other units' equipment and procedures were improved. My heart goes out to Ron's family and shipmates.

Lessons Learned:

1. Changing tactics caught our adversaries literally sleeping. Thanks to Alexander Hamilton.
2. It's more stressful ordering your troops on a dangerous mission than going on it yourself.
3. Sometimes big gambles pay off.
4. It's tempting to celebrate after a big victory, but it's better to exploit the situation quietly.
5. If you're going to conduct a middle of the night, unannounced boarding on a ship smuggling tons of drugs, take insolated gloves and dog biscuits.
6. Always look for ways to buy down risk and ask what-if questions. This is better done in person on the water than from behind a desk.
7. Case package and paperwork are boring but critical in convictions. Almost fifteen years later, cartel leader El Chapo was on trial, and our boarding officer testified based on previous notes.

Chapter 24

Semi-submersible, Fast Roping and Shooting From Helicopters

WE WERE HAVING A good run of it with drug busts, but it's not always as simple as finding a huge number of bales on a slow-moving fishing boat. Smugglers rapidly change tactics—but we can too! The Coast Guard created a special helicopter unit called HITRON, which became highly effective in stopping high-speed craft. These armed helicopters were designed to fly from Coast Guard cutters and fire warning shots by machine gun and then precision fire from a fifty-caliber sniper rifle into the outboards. Frankly, there were more drug boats (we nicknamed go-fasts) to prosecute than the Coast Guard, Navy, or allies had assets to intercept. The TACLETs did not have an embedded aerial use of force capability at the time, meaning US Navy and Allied ships were losing potential bust opportunities. Both TACLETs recommended this capability be created quickly to counter the smugglers' shift to small high-speed vessels.

What sealed the deal was my trip to Key West to visit the Joint Interagency Task Force South. This internationally staffed unit under US Southern Command's (USSOUTHCOM) was charged with managing the Department of Defense and Allied efforts to stem the drugs heading out of Central and South America. Who was in charge? None other than my old boss RADM Hathaway.

In route to Key West, I got word of a PACTACLET LEDET aboard a Navy frigate involved with a high-speed surface chase. They fired warning shots and disabling fire. Because the Coast Guard couldn't shoot from a Navy helicopter, we followed existing policy and had a Navy sharpshooter sandbagged up in the mast of the ship. Obviously, we'd rather shoot down than horizontally to prevent skipping or collateral damage to another vessel.

The vessel didn't comply with the warning shots across their bow, so the sharpshooter fired on the outboards. Eventually, the chase ended. When the LEDET boarded the adrift vessel, they found carnage. A man on the back deck had the back of his head blow off, and another man had a huge round through his leg. After closer evaluation, the deceased man committed suicide and was probably the drug trafficking organization representative and responsible for the load's safe delivery. The leg injury was likely a ricochet off the outboard.

Pointing to this case, we sent select members of LEDETs to sniper school and started the airborne use-of-force program for TACLETs. We secured a range at Camp Pendleton and had the Navy Helicopter pilots certify with our gunners before deployments. This led to more seizures.

Just as you counter smugglers with a new interdiction capability, they come up with another method of smuggling. In this case, our surveillance aircraft overflew a trawler with long cables extending into the water. It was deep there, so cables in the water was unusual. The ship with the LEDET intercepted. At the time of the boarding, the cables had been pulled in and showed signs they had been cut, and the crew was acting suspiciously. The ship and aircraft backtracked where the vessel had come from, and sure enough, they located a floating object that looked like a large metal floating cylinder. It had a cut tow cable that matched

First drug-laden submersible

the trawler. It had fins and was designed to submerge when it was being towed. Even better, it was full of cocaine. PACTACLET seized the first of many submersibles to be captured.

Fast forward fifteen years. You may have seen in the news the clip of a PACTACLET chief jumping on a manned speeding semi-submersible. If you watch that video closely, you'll see his partner spraying pepper spray into the air intake. Better to force them out rather than having to

PACTACLET chief boarding a submersible (photo courtesy of DVIDS)

go in and get them. The origin of these submersibles was linked to the Tamil Tigers engineers that fled Sri Lanka. I had heard all about them during my month in India.

At this point in my career, I learned the best boarding officers are tenacious. On one occasion, a LEDET boarded a large purse seining vessel that uses a small boat to encircle tuna with a net. The net itself had yellow floats on the top and weights on the bottom. The net is then drawn up and pulled alongside, hopefully, full of tuna. We had information that the vessel was possibly involved in smuggling, and the crew was acting a little suspicious.

We have a very expensive piece of equipment that the LEDETs carry with them on deployment called an ion scan machine. You may have even seen them in the airports used by TSA. They can be calibrated to detect small quantities of explosives or drugs. Because it's electronic, very bulky, and one boat ride could destroy it, this machine is left aboard the host naval vessel in a controlled environment.

The boarding team goes aboard, and if there is suspicion of smuggling, the team uses cloth swaps throughout the vessel, noting their locations, and then run those back to the naval vessel and through the detector. On this particular purse seining vessel, the team got hit after hit for cocaine, but the search efforts over many hours resulted in no unaccounted for space that could contain drugs. After four to five hours, the fishermen, Navy ship, and the boarding team were getting frustrated and knew they had to depart soon. It was possible the crew had offloaded before the boarding team arrived. Just before concluding the boarding, one of the team members felt one of the buoys on the net. Having done fisheries boardings before, he thought it seemed a bit heavy. He pursued permission for a destructive search. Sure enough, all the floats contained a kilo of coke inside. They seized the boat and arrested the crew. The authorities in Ecuador raided the fish float factory.

In another such case showcasing LEDET tenacity, a boarding team was on a large fishing boat and got ion scan hits for cocaine. Again, it had taken several hours, and the team's search found nothing. But the boarding team recalled a different suspect vessel had been towed in for further inspection pier side a couple of months earlier. When that suspect vessel arrived in port, the suspect crew had pumped their fuel tanks dry during the middle of the night. No drugs were found on that vessel. On a hunch, the boarding team used a "theft sampler," an instrument designed to capture a small sample of liquid from the bottom of a fuel tank. When they pulled the sample out and dried it, they were able to convert it to crack which then tested positive for cocaine. The tank proved to be full of cocaine in liquid form—this was the first maritime interdiction of liquid cocaine.

Self propelled semi-submersible SPSS

Our tactical efforts continued to expand based on emerging threats and the dramatic increase of active shooters. In decades past, when there was a shooter, our tactic was to surround the suspect and contain their movement until a police negotiator arrived. The new active shooter that we've unfortunately seen arise is motivated by killing as many people as they can in as short of time as possible. This changed the law enforcement tactics, encouraging immediate engagement with the threat. More often than not, the shooter will panic, their gun will jam, or the shooter will be neutralized or commit suicide. Law enforcement ashore generally has a very fast response, usually within minutes. Some of the worse active shooter events are when, for whatever reason, law enforcement is delayed or slow to respond.

The maritime environment, by its very nature, will have a delayed response. Boats are slower, equipment is different, things on vessels can explode or the vessel can sink. If active shooters control the vessel, whether it be a ferry, cruise ship or smaller passenger vessel, they will not be assisting a responding law enforcement. They will likely repel boarders trying to retake control of the vessel.

The Coast Guard had to develop a capability that can be rapidly deployed. We refined the use of caving ladders, which are metal-cable, rope ladders with rungs the width of a person's foot with an extension pole with a hook on the end. Think about storming a castle in the old days; that's what this will look like except the boarding team will be climbing up from a delivery boat. This will take time, so we also coupled that with fast roping from our helicopters.

Essentially, a team of six to eight people, tactically outfitted, hover over the vessel that has been taken over. They drop a heavy, two-inch diameter rope from the helicopter's open door to the deck

A "stick" or boarding team of eight entering an HH60 helicopter for fast rope training (photo courtesy of DVIDS)

of the ship. Then the team grips the rope at chest level wearing welder gloves and squeezes the rope between their feet. Once leaving the door, gravity does the rest. As a forty-year-old commander, I must admit this rivaled my early days of sky diving.

Sliding down a rope in the middle of the night to a ship with armed people who want to shoot you is quite unnerving, but it's better than doing nothing if we have an active gunman on a cruise ship or a ferry. That's why we built and exercised this capability know as a HAF/BAF (helicopter and boat assault force). We practice alongside the FBI's hostage rescue team (HRT) and the Navy Sea, Air, and Land (SEAL) teams because the bigger the force, the more likely it will prevail.

Lessons Learned:

If you can dream it, smugglers are probably already doing it.

Never go into a boat or building owned by hostile inhabitants if you have a choice. It might be better to flush them out with pepper spray or tear gas or just wait them out.

If we chose not to do this mission, who can or will?

Coasties can shoot out of Navy and allied helicopters.

Caving ladders are dangerous but need to be exercised regularly. Use a safety line.

Fast roping from a helicopter at night is very scary, especially if it's to a small floating target with people that want to shoot you.

Chapter 25

The Dutch, a Seized Freighter and Hurricane Katrina

HURRICANE KATRINA WAS A disaster for the country and specifically for my parents' original hometown of New Orleans. I had seen firsthand what Hugo did to a small town in St. Croix. There was no comparison. The Coast Guard poured in resources, and in those first few days, they had been phenomenal. Out in San Diego, we heard that things were starting to deteriorate on the ground and some of the rescuers were now taking fire from people in the flooded city. There was a call for armed boarding officers to accompany the ground rescue teams doing the house-to-house searches.

We had a LEDET in the Caribbean operating from a Dutch ship which had just seized a coastal freighter with cocaine. The Dutch command watched the news of New Orleans unfold on satellite TV. The initial reports stated looters and violence were hampering critical supplies and rescues. There was a total breakdown of law and order. I'd seen this movie before. I knew PACTACLET could assist.

The situation continued to deteriorate, and the Federal Emergency Management Agency (FEMA) director was a bit overwhelmed. He was a political appointee and probably didn't have an emergency management background. Honestly, only a handful of people did, and this would have challenged anyone. I won't get into what level of government should be responsible for what or where failures occurred in this unprecedented storm response. Still, it was determined that the Coast Guard's swift actions saved roughly 33,000 people. The service was even named ABC's

People of the Year for their efforts. VADM Thad Allen was handpicked by the Bush Administration to take over as the National Incident Commander after some initial management concerns. VADM Allen's

chief of staff was Captain Tom Atkin.[1] One thing became obvious to these leaders: many forces wanted to assist, but they were not self-sustaining or effectively organized.

In some cases, they exacerbated the problems already in the region. In fact, they required the crippled existing command to set up support logistics for them. That's VADM Allen made it

President George W. Bush with Admiral Thad Allen at the Coast Guard Academy after Hurricane Katrina (photo courtesy of DVIDS)

a priority during his later term as Commandant to create self-supported deployers to come to the aid of resident-impacted commands.

Meanwhile, the Dutch had pulled into Puerto Rico to transfer custody of the seized freighter, detainees and cocaine. Then eager to support the New Orleans recovery effort, the Dutch ship loaded a second helicopter and flood punts (shallow water rescue boats) and made best speed to support hurricane relief efforts on the gulf coast. Their ship was assigned to Mississippi's response. I engaged with those Coast Guard units responding in New Orleans who stated they needed our LEDET, which was aboard the Dutch ship, to support the rescue teams going into the city doing welfare checks. The problem was the Dutch didn't want to let go of the LEDET as it was their pass to get in and support the broader US hurricane response.

While I certainly respected that, the LEDET had a higher tasking. I had to call Amsterdam and eventually get the guys freed up to respond in New Orleans. There, along with other members from PACTACLET, they saw horrific sites in the door-to-door checks. I will spare you the graphic nature, but understand starving pets, snakes and alligators were everywhere—as were human casualties. The floodwater was indiscriminate on the lives it took. Luckily, several of our members were

[1] I later worked for Tom Atkin at the US Deployable Operations Group

trained as counselors for those who experience traumatic events. They were very much employed.

I had an Academy classmate named Scott Hitchen. He was a big, happy-go-lucky guy and a bit of a party animal. He went into the aviation field, so I didn't have much contact with him after graduation. He was a lead pilot during Katrina. The Coast Guard moved heaven and earth to get as many helicopters on the scene as possible, and they were working nonstop. Scott would never offer this up, but he was named officer of the year for his heroics during Katrina. I saw him years later. He was a completely different man than I knew at the Academy. I can testify that he was still big, because he jumped on my back. He was now happily married and in the process of adopting orphans from China. He was certainly not the only saint I ran across during my career, but that's what they called him in New Orleans.

Fifteen years later, on a trip to New Orleans with my daughter to see a Saints game, we saw monuments and graffiti paintings, thanking the Coast Guard for rescuing their family members.

Lessons Learned:

1. Get your guys into the action.
2. Natural disasters bring out the best in people initially and then the worst if it drags on.
3. Get people counseling if they witness horrific things—even if they say they are fine.
4. Working with allies can be tricky if our motivations differ—and they do.
5. It's nice to be reminded years later of your service's successful missions.

Chapter 26

Iranian Aggression at the Iraqi Oil Platforms

I MADE SIX TRIPS to the Middle East; three while on active duty and three as a civilian trainer. In the first two, I sustained injuries, but it wasn't the Purple Heart kind.

I must admit, when I sailed across the Atlantic for Desert Storm, I felt a bit cheated that we were dumped off at Rota, Spain. I know that seems wrong, but I was single, invincible and in my twenties. Now at almost forty, it was time to complete unfinished business.

My first trip to the Middle East was shortly after taking over the TACLET. Months earlier, TACLET South's Petty Officer Nate Bruckenthal and two Navy sailors were killed in action when boarding a dhow rigged as a floating IED. TACLET's Petty Officer Joseph Ruggiero was also injured in the blast, along with three other sailors. Joe, injured in the water, managed to inflate his camelback and stay afloat until rescued. They all received Purple Hearts.

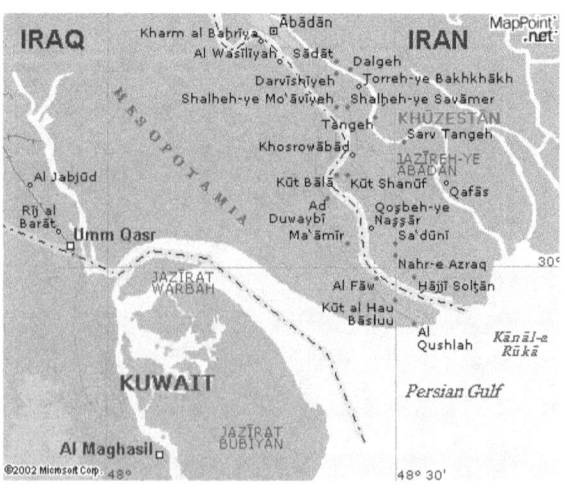

The TACLET was operating from the Navy's USS *Firebolt*. Their mission was simple: protect the two strategic oil platforms KAAOT and ABOT, located in the Persian Gulf just south of Iraq but adjacent to Iranian waters. These oil platforms were responsible for over 90 percent of the oil coming out of Iraq.

The revenue generated from this infrastructure was critical to rebuilding Iraq after the war. That made protecting these platforms critical in the United States exit strategy. The LEDET's role was not only boardings but training Iraqi Marines to take over this responsibility. The LEDETs were either operating from Bahrain on small US Navy gunboats or in Umm Qasr on the ground in Iraq. My role as the new TACLET CO was to ensure deploying LEDETs were professionally trained and equipped for what had become a deadly mission. Once they were trained and equipped, they traveled to the Middle East and fell under the local commander's tactical control.

To do my job right, I needed to meet with those in tactical command of the missions based in the Middle East. The best way to protect KAAOT and ABOT was to mount stand-off weapons on the platforms and place a ring of floating steel around them to intercept threats.

US Navy Typhoon class patrol boat, my ride around the Iraq platforms (photo courtesy of DVIDS)

That is exactly what Nate and Joe did. They completed the mission but at a huge cost.

As for the ring of steel, the Coast Guard sent six, 110-foot, patrol boats (like the *Key Largo*) and an overarching command known as Patrol Forces Southwest Asia or (PATFORSWA). An interesting coincidence, the XO of PATFORSWA on my first visit was LCDR Laura Dickey. Laura had been my first XO on the *Key Largo*. Fast forward a decade and my second XO on *Key Largo*, Lee Petty, had promoted to captain and eventually commanded PATFORSWA.

Thankfully, when the Coast Guard stood up the Middle East unit, they did not take the boats from Puerto Rico; they were sourced from around the country. My experience with the 110s and now the boarding teams and a Port Security Unit gave me an advantage in evaluating the use of these assets in theater. I was no stranger to long international flights at this point, but this seemed particularly grueling as I experienced some health issues on the plane.

When we finally landed in Bahrain, it was a balmy 114 degrees. I'm not telling military vets anything new. But for those few who have not been to the sandbox, it was like a hairdryer blowing in your face with some sand sprinkled in for good measure.

The LEDET in Bahrain had been deployed for two months, so they were the last PACTACLET members I had not yet met. The last thing I wanted to do is roll into the combat zone with a medical issue, but that's exactly what happened. I met up with the deployed team, toured their shoreside accommodations, met the PATFORSWA Commander also referred to as commodore since he had multiple ships working for him and of course Laura. One of the coolest things

After six trips to the Middle East, I got to ride a camel.

to do in Bahrain is to drive their really fast go-carts. Certainly was a good time to mix with the crew, but I was also doing my best to control some internal pain. Then they had arranged a flight out to the USS *Thunderbolt*, which was a Navy Cyclone Class patrol ship or Patrol Coastal. This was the same class ship doing the same mission as the USS *Firebolt* with Nate and Joe. USS *Thunderbolt* was underway off KAAOT, so we needed to get there.

The flight was in the "Desert Duck," the nickname for an ancient helicopter that sprayed hydraulic fluid on my desert camouflage uniform. The flight mechanic said the only time it becomes a problem is if it stops spraying. We flew to an Australian cruiser serving as the Allied Command and Control platform for the mission and then transported by small boat to the much smaller USS *Thunderbolt*.

I berthed in the aft SEAL berthing compartment with the rest of the LEDET. Our guys were doing well and had been involved in interdicting a vessel with mines aboard. *Thunderbolt's* skipper was a Navy

The "Desert Duck." My ride to USS Thunderbolt

LCDR, and by all accounts, he could have easily been a Coast Guard patrol boat skipper.

Having the Coast Guard boarding teams filling the protection mission, the SEALs moved ashore to go after targets up in the mountains. In addition to manning the navy ships, another LEDET was staged in Umm Qsar from a relatively small British compound where they trained Iraqi marines.[1] PACTACLET and TACLET South (Miami) would alternate duty sites.

Reprovisioning the small British base was left up to its inhabitants. That meant running a small convoy force through the desert to a supply depot. With the small base, our LEDET personnel were some of the best trained for providing convoy security. Plus, they were consuming supplies. The LEDET let me know that they were requested to go on these convoy missions. When I advised our Coast Guard Headquarters, still reeling from Nate's combat death, they denied the request. This put our folks in a difficult position. I lobbied for and got some specialized convoy training and eventually received permission for them to go on the supply runs. (I'm pretty sure they were doing it anyway.)

So, I mentioned I was feeling ill, and the condition worsened. I don't often get sick or suffer an injury, but unfortunately, this was one of the times. I could no longer disguise it. I had shooting pain, could barely stand and had a fever and sweat coming on. There was no doctor on the *Firebolt*—it was only slightly bigger than the *Key Largo*—and I didn't want the Navy cook cutting on me. It turned out to be kidney stones.

---

[1] About three months later, in November 2004, the greater Allied land force moved to Fallujah, and our folks were no longer in the rear with the gear. Their camp took fire but did not return it due to possible collateral damage.

It's bad enough if it happens at home or perhaps in a US hospital. But I was curled up in a SEAL compartment of a rocking Navy gunboat doing circles around the Iraqi oil platform which had been attacked four months earlier.

Just when I thought it couldn't get worse, the ship went to general quarters or battle stations. The Iranian gunboats love to taunt the United States at every opportunity, and they were running their gunboats right alongside our vessel. Eventually, it passed . . . both the Iranians and the kidney stones.

I also was taking satellite calls during this time from the other side of the world. Our teams in the Pacific were having great success. I was on the bow of the *Thunderbolt* when Rod Rojas briefed me on the successful UNB of the *Lina Marie*. I think those calls and a whole bunch of water improved my condition.

But it was my second trip to the Middle East that provided the biggest scare.

On this trip, still in command of the PACTACLET, I went with my boss from the Pacific Area, CDR Brian Thompson. I chugged water and cherry juice on the way over, more than I ever had in my life. Rank-wise, I was senior to Brian, but positionally, he was my boss. He was awesome. He was always a voice of reason and support in the Pacific Area, which is normally dominated by major cutter captains. Brian was the Special Forces guy. Also with us on the trip was LT Scott Toves, the new XO from TACLET South.

We flew into Bahrain jet-lagged but kidney-stone free. I learned from my last trip that if I didn't get the blood flowing with a workout, I could face round two, and I wasn't ready for that. So, at approximately two a.m. local time, I was lying there wide awake in the hotel room. I got up and explored the hotel. There was a fitness center, but when I tried to open it, I set off an alarm, so I quickly departed. On the way back to my room, I saw the hotel had a nice outdoor swimming pool one level down. Despite being the middle of the night, it was probably in the low nineties. I grabbed my bathing suit and a towel and returned to the concrete staircase that spiraled down to the pool deck.

A sign in Arabic blocked part of the staircase, but looking around it, I saw a guy cleaning the pool with one of those extended nets. So, I slipped around the sign and headed down the stairs, when suddenly the concrete turned gravel, and I was free-falling until my face gained traction on concrete and a rebar. Surprisingly, it didn't hurt too badly, and I pulled myself out of the hole, brushed off and continued to the pool, but my head was bleeding pretty good.

I used the towel for direct pressure, made it to the guy and asked if I could jump in.

The pool guy was probably used to doing his duties in absolute solitude, and my arrival must have startled him. He took one look at my head and the bloody towel and shook his head, pointing at mine. I'm pretty sure he didn't speak much English, but I asked if maybe the front desk had a bandage.

I tried to do a patch job and figured swimming was out, so I headed back to the room. After another thirty minutes of direct pressure, I thought I better call my traveling companion Scott. I don't think he was sleeping well either, so he came to check on me. By this time, the white towel was more red than white. I suggested that we play a trick on Brian. It was perhaps three a.m. I slumped over Scott and knocked rapidly on Brian's door. As he opened the door, there I was, a big bloody mess.

I said I had gone running and heard gunshots, so I fell to the ground. I claimed I wasn't sure what happened, but I might need to go to the hospital. I purposely slurred some of my words.

Brian jumped into action and dialed 911, which, I might be wrong, probably did not work in Bahrain. Just before he connected, I stood up and said, "Surprise! It's not that bad."

He probably wanted to shoot me, but we had a fun story to tell. I did have to go to the hospital to get cleaned up with a nice bandage.

My travel mates gave me the nickname "Head Wound Harry" for the rest of the trip. When I went onto the US base for chow, I received a thumbs up from the troops. No time to explain what truly happened!

On the way home, I called ahead and explained to Nye that I had gotten a slight injury, which is not what a family wants to hear from a spouse returning from the Middle East. But in the end, I ripped off a sideburn, and I have a scar as a souvenir.

Lessons Learned:

Don't go around signs blocking stairwells.

Head wounds don't hurt much, but they bleed a lot.

As an additional benefit, a head wound provides a handy excuse when you can't remember names or when you have other senior moments; attribute those moments to a "war-time" injury you don't want to talk about.

Sometimes a Band-Aid just won't cut it.

Make sure before you pull a prank on your boss, he has a sense of humor.

Construction signs at a hotel should be in multiple languages.

Kidney stones are perhaps the worse pain a man can feel. Hydrate often, especially if its 114 degrees.

Chapter 27

Capture of America and Mexico's Most Wanted

THE DEA'S SAN DIEGO Field Office initiated Operation Shadow Game to apprehend the Arellano-Felix organization (AFO). Also known as the Tijuana Cartel, the AFO has moved billions of dollars' worth of Mexican and Colombian drugs into the United States while committing some of the most vicious murders seen in the drug underworld. This organization controlled a third or more of the cocaine traffic entering the United States and spent countless millions on buying protection from Mexican police, judges and military officials.

Francisco Javier Arellano Felix was among eleven people charged in absentia in 2002 with several conspiracy and racketeering counts. The indictment linked the AFO to a 1996 killing in Coronado, California, and the 1992 shoot-out at a discotheque in Puerto Vallarta, Mexico. Javier Arellano is suspected of conspiring to assassinate Roman Catholic Cardinal Juan Jesus Posidas Ocampo, who died in the Guadalajara International Airport parking lot after receiving fourteen gunshot wounds.

Over the past fifteen years, several of the leaders have been captured or killed, leaving the youngest brother, Javier, in charge of the day-to-day operations of the AFO. Known as El Tigrillo, or the Little Tiger, for his violent nature, he was one of the United States and Mexico's top-most-wanted fugitives, leading the US State Department to offer a five-million-dollar reward for information leading to his apprehension and capture.

In late 2005, the AFO purchased a US-registered, sport-fishing boat, the *Dock Holiday*, in southern California. It was moved to La Paz, Mexico, for use by leading members of the organization, who were avid deep-sea fishermen. The concept of Operation Shadow Game was simple: after receiving specific targeting information and intelligence from the DEA, prepositioned, specially trained, Coast Guard, waterborne assets would intercept and board the *Dock Holiday* in international waters, remove the DEA designated suspects and deliver them to the United States.

Early in May 2006, the DEA approached the Coast Guard through the Coast Guard Investigative Service. On the twelfth of May, the Eleventh District hosted an operational meeting with DEA to discuss the concept and clarify the details of the investigation, potential threats to boarding teams, expectations from the Coast Guard, authorities and jurisdiction. On the thirteenth of June, the DEA in San Diego hosted a tactical planning meeting with district's enforcement, intelligence and legal staff and PACTACLET to discuss placement of at-sea units, interdiction tactics and disposition procedures. Consideration was given to the violent nature of the AFO. Given their historical involvement with killings of Mexican authorities, their employment of violent gang members as bodyguards, their possession and access to armaments to include automatic weapons and grenades, the planning group assumed that if the *Dock Holiday* was interdicted at sea, those on board would forcibly resist.

At the meeting, I recommended that specialized, prior trained TACLET or MSST boarding teams were best equipped to execute the boarding based on close quarter combat training. Due to the extremely sensitive nature of the operation, knowledge and planning were limited to a small cadre. I was blessed with a great XO named Ed Songer and Command Master Chief Ted Fuller. These two guys were great advisors and trusted leaders, who allowed me to travel down range and to the Middle East frequently. But this was probably the highest risk evolution we ever attempted. We were just a small part, as this was being led out of the Eleventh District.

This multi-unit team that included PACTACLET developed the at-sea portion of Operation Shadow Game. Working with the district enforcement staff, the legal team provided advice throughout the operational planning stage, identifying potential statutory and international pitfalls and refining the plan to ensure absolute compliance with legal requirements. To ensure alignment with senior Coast Guard leadership, the district staff briefed the Commandant's staff and continued to provide updates throughout the plan's development.

The mission took tactical control of cutters *Steadfast, Zephyr* and *Monsoon* (which were the same class as USS *Thunderbolt* but now were Coast Guard cutters), the *Edisto*, which was a 110-foot patrol boat like *Key Largo*, and three, eight-person, noncompliant-vessel boarding teams from PACTACLET, MSST 91109 (San Diego) and MSST 91103 (Los Angeles). The specialized, handpicked, boarding teams were designated to deploy aboard assigned cutters in anticipation of executing the boarding. The plan developed by the district provided a continuous, in-theatre coverage from July 9 through August 10.

Additional planning and coordination by the Eleventh District staff—led by Brian Thompson—ensured each force package was properly trained, outfitted and ready for operations. Due to DEA resource availability, Operation Shadow Game was scheduled for completion on the tenth of August.

One of the most critical phases of preparation involved at-sea training evolutions with cutters *Monsoon, Zephyr* and *Edisto*. The district obtained a towable target vessel from CBP's seized asset inventory and oversaw training for the cutter's sharpshooters and gunners to execute tactics against a noncompliant vessel.

PACTACLET training staff conducted four weeks of focused training in noncompliant boarding, opposed boarding and extraction techniques to prepare LEDET 108 for the operation. I had pulled them out of the deployment schedule, and along with the training staff, we compartmentalized this mission, not sharing details with the other LEDETs. We even had the *Dock Holiday's* schematics and did tape drills and worked with the SEALS in San Diego to get extra equipment and secure training sites.

Steadfast, Edisto, embarked aircrew, and LEDET 108 conducted drills and exercises to integrate the entire force package underway. We prepared for the possibility of a combined boarding. We also provided an additional two weeks of extensive training to CGC *Monsoon's* boarding team in the finer points of close quarter combat, combat first aid, advanced weapons training, combat marksmanship, operational risk module application, weapons retention, contact weapons defense, simulated munitions training and other team dynamics.

Mark Ogle

Prior to deploying aboard the *Monsoon*, MSST 91103 (LA/LB) made extensive preparations for their participation in the operation. Immediately upon embarking on the *Monsoon*, MSST 91103 Noncompliant Boarding Team (NCBT) members integrated with the crew and conducted extensive preplanning briefs.

Daily training sessions, walkthroughs and at-sea exercises ensured complete force packages unity and readiness. This included working through scenarios for any contingency, including the possible presence of children aboard the suspect vessel.

Monsoon. A Typhoon class Navy ship was now a Coast Guard Cutter (photo courtesy of DVIDS)

Because the operation was scheduled to conclude August 10, most of the MSST and NCBT disembarked during the *Monsoon's* port visit in Mazatlán on August 11. So much effort had gone into this mission and not having an arrest as the window closed was extremely deflating to all involved. But wait!

On August 14, two days after departing Mazatlán, the district advised *Monsoon* of a *Dock Holiday* movement. *Monsoon*, now with essentially just their crew and no backup, altered course toward the last known position of *Dock Holiday*. The Eleventh District directed *Monsoon* to close to a position just over the horizon from the vessel and provide a risk assessment of their ability to stop and board the *Dock Holiday*. The cutter and MSST CO scored the mission as high-risk yet expressed confidence in their crew's ability to stop and board *Dock Holiday* given these conditions.

The Eleventh District commander authorized the *Monsoon* to immediately proceed with the intercept and board *Dock Holiday* before they began to trek toward Mexican territorial waters. Due to superior employment and placement of the cutter and small boat, *Monsoon's* six-man boarding team arrived alongside the *Dock Holiday*, embarked the vessel and rapidly established positive control of the situation. *Monsoon* took digital photos of all eleven personnel aboard *Dock Holiday* (eight adults and three juveniles), which were relayed to the DEA through the Eleventh District Command Center. The boarding team also transmitted scanned images of the fingerprints on all adults being held by *Monsoon*. We had a match!

At the same time, coordination was taking place for the disposition of the vessel, eight adults and three children. *Monsoon* transferred all eleven to the cutter, embarked a custody crew aboard the *Dock Holiday* and quickly proceeded further offshore to commence their transit to San Diego. During the voyage, the *Monsoon* refueled *Dock Holiday* several times, eliminating the need to take the vessel into tow for the lengthy trip to San Diego. Throughout the return transit, *Monsoon* ensured proper care and custody of the eight adults and three children. That included allowing the parents to spend time with the children. This was appreciated by those detained, and the two cartel leaders eventually revealed to the Coast Guard crew their true identity.

Monsoon traveled north alongside the Baja peninsula, returning to the United States. For extra security, the district directed the 378-foot cutter *Boutwell* to intercept *Monsoon* and provide escort and refueling services for the remainder of the transit. *Boutwell* arrived on the scene with *Monsoon* on August 15.

Respecting partner agency desires to commence questioning of the detainees as soon as possible, the district coordinated the transport by helicopter of three personnel (one member from DEA, FBI and US Attorney's office) to the ship. The *Monsoon* and her escorts arrived in San Diego on the morning of August 17, as scheduled.

Following Mexico's press release on August 16, the Commandant of the Coast Guard, DEA Assistant Administrator for Operations, and Deputy Attorney General held a press conference in Washington DC during which Mr. Michael Braun from DEA said, "We would not be here today if it were not for the outstanding support

Turning over the cartel leader in San Diego

rendered by the US Coast Guard. . . . It is a testament to this dynamic, very flexible, maritime force. . . . The flexibility that is needed as our nation continues to wage war against drugs and terrorism."

DEA Administrator Karen P. Tandy summarized the impact of Operation Shadow Game in a press release on August 24:

> For over a decade, the Arellano-Felix family dominated the Mexican drug trade and flooded our nation with hundreds of tons of cocaine and marijuana, and massive quantities of methamphetamine and heroin. Javier Arellano-Felix, the last stronghold at the top of that cartel, is a violent drug kingpin wanted in the United States for numerous drug trafficking, conspiracy, and money laundering charges. He is considered threatening enough to our nation to warrant a $5 million US Department of State reward for his capture. His arrest topples a dynasty built on violence and drugs and puts a chokehold on the destruction this brutal organization has caused in both the United States and Mexico.

In the two weeks following the takedown, Mexican authorities arrested several dozen cartel members and crooked Mexican law enforcement officials. The bloody feuding between competing drug smuggling factions resulted in the deaths and assassinations of several other drug cartel members.

During that last year I was TACLET, we received many senior visitors. The United States Interdiction Coordinator gave us the national award for intelligence. Admiral Allen, now Commandant, and VADM Dave Pekoske, the new Pacific Area Commander, were among them. We also had congressional visitors there for briefs to see what we were doing. The newly promoted RDML Tom Atkin was on one of these visits and pulled me aside and asked if I'd like to be the first operations chief for Coast Guard Special Forces. Remembering the last time I was asked such a question from a one star—RDML Harvey Johnson—I obviously said yes. This time, however, I actually felt better prepared for the job.

There was a final visitor who made his rounds through the Coast Guard; Dave Helvarg was writing a book. Much like the visit I had years earlier aboard the *Vashon* who chronicled the St. Croix Rescue, this book looked across the service to include exciting stories covering numerous missions. I gave him a tour and made sure he met the young men and women doing these dangerous missions. Sure enough, TACLET and the new Special Forces initiative made it into his book, *Rescue Warriors: The U.S. Coast Guard, America's Forgotten Heroes.*

Lessons Learned:

Cutting the head off the snake may be a good idea, but it has consequences.

Getting out good news early can create bad news quickly. The offload of the cartel boss was supposed to be kept quiet, reducing the chance of a cartel rescue mission. Unfortunately, overeager public affairs folks leaked the story. PACTACLET sent armed reinforcements to assist DEA and the US Marshalls upon arrival of *Monsoon* in San Diego

If there are children present, this can be to your advantage in keeping the situation calm.

Going after the most violent elements will discourage others from using violence.

Delaying departure on a stakeout, as *Monsoon* did, for just a few hours can make all the difference in the world.

Chapter 28

Coast Guard Stands Up Special Forces

THE COOLEST THING ABOUT making O-6 or captain is getting the eagle sticker for your car that allows you to park closer to the exchange or commissary. Plus, it's great for a book title.

Admiral Allen was waiting for this moment. He had several initiatives he wanted to implement upon taking over as Commandant. Tom Atkin had been his chief of staff during Hurricane Katrina and was recently promoted to flag. RDML Atkin was selected to command the new United States Deployable Operations Group or—as Admiral Allen named it—the DOG.

RDML Tom Atkin in the Middle East with Coast Guard Specialized Forces

RDML Atkin was only four years ahead of me and two years ahead of the new deputy, Captain Gary Rasicot, so he was looking for a boot captain to serve as his operations chief. I was as boot as they got. I had to be frocked, meaning I got to wear the rank, but I was still getting paid as a commander.

My previous carpool mate Keith Smith and I had been in command of units in southern California, and we had an unwritten agreement that we'd try to go to this new unit together. RDML Atkin had been a TACLET commanding officer for Keith, and because RDML Atkin really wanted Keith, I got in on the package deal! I also think he wanted someone with a similar mindset that was operationally aggressive in the position. I thought I was operationally aggressive until I met RDML Atkin.

PACTACLET had chalked up rather good stats in San Diego, and my time in the Port Security Unit and IMLETT also served me well in the new position. There were 3,000 people in twenty-seven units under the new structure, and pretty much all of them fell under operations. That's a lot of evaluations! It's one thing to oversee something like this; it's quite another to build it from scratch while simultaneously personally deploying on missions. I was given four days to drive across the country from San Diego and report into my office in DC. I was immediately deployed on a mission. I returned from that mission in Miami just in time for the DOG commissioning ceremony. Fran Townsend was in the audience. She was now in the big league as the White House Homeland Security Advisor. I knew her when she headed up the Intelligence Branch at Coast Guard Headquarters and had frequent meetings with Admiral Johnson.

After some quick speeches, it was time to cut the cake. Somebody handed me a bayonet. As I made my way to deliver the cutting tool to the boss, I came too close to Fran Townsend. Out of nowhere, I was intercepted by three agents with ear mikes.

One said, "Hold up, sir. You got red dots on you." I gladly passed the bayonet off as I had lost my appetite.

The DOG created adaptive force packages to deliver what a supported command needed in the most efficient way. Deployable elements were assigned within the DOG staff in Washington, DC. So, if there was a unique need for, let's say boat security and pollution response, the deployable element filled the leadership role from multiple elements in different special forces units. That leadership element supported the tactical commander. The tactical commander could be a cutter, sector, district, or combatant commander (COCOM). I stood that recall status 50 percent of the time and was frequently deployed.

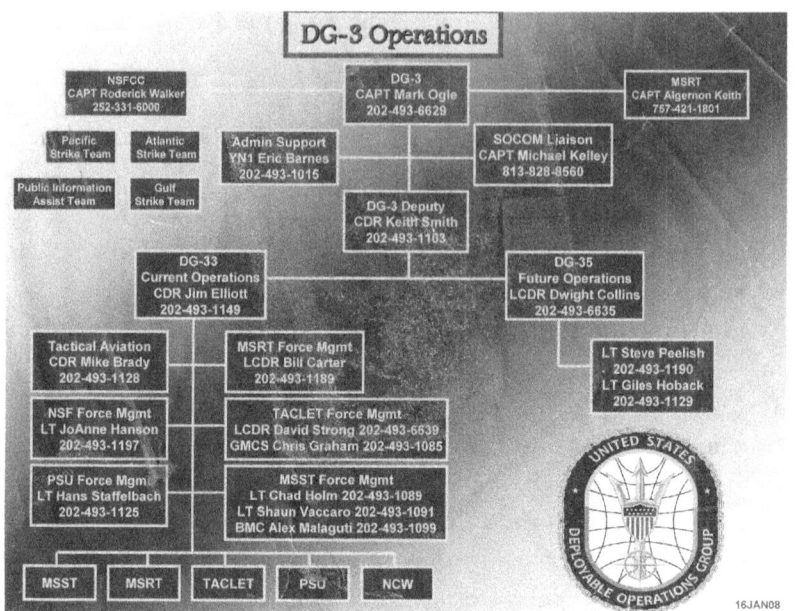

The organizational chart for Operations within the DOG, composed of 3000 members

I thought my time as executive assistant for Admiral Harvey Johnson was demanding; this was an even bigger test. Admittedly, I had no great desire to go back to Washington, DC, but it was an opportunity of a lifetime. Plus, it was not actually in our headquarters building; it was in Ballston next to a mall.

Many of us in the Coast Guard wanted to build up Special Forces, but there had always been a resistant culture. As we picked up the banner, we were mowed down. But this was different; the four-star wanted to make it happen.

Admiral Allen involved the detailers and stacked the deck for the new unit. Keith Smith was my deputy, and our current operations officer was Jim Elliot. Both were amazing and had been recognized as officers of the year. My counterpart and good friend Merrie Austin ran the planning section. Tom Allen was the finance guy.[1] The second year, Jim Tunstall, my former detailer, joined us as the O-6 liaison to Special Operations Command (SOCOM). If that wasn't enough, we had a Navy Seal captain and former school chief for BUD/S on the command staff.

---

[1] Merrie, Keith and Tom all made flag officer. Gary Rasicot entered the Senior Executive Service as a civilian equivalent to an admiral. Admiral Atkin became Acting Assistant Secretary of Defense for Homeland Defense.

The DOG sent and graduated Coast Guard members from SEAL training. RDML Atkin and I met with MARSOC, the Marine Corps Special Ops branch, and Special Operation, Southern, Central and Northern Combatant Commands. We were really creating momentum for the new organization.

*Coast Guard members attended BUDS
and earned SEAL pins*

Lessons Learned:

If you're going to stand up Special Forces, you better have the blessing of the top DOG.

Sourcing a new command with no additional personnel coming into the Coast Guard means other units will have to give up personnel. That can create friction for the new unit.

Managing 3,000 operators on global missions while personally deploying in a port starboard rotation is very demanding.

RDML Atkin was a busy guy. He made it clear we had no spare time to waste. Senior staff could discuss issues amongst themselves, so we weren't running after different rocks. That said, he put a chin-up bar outside the door. If you wanted to speak to him, you had to do a pull up.

Getting an all-star team is critical. Unlike the other services, the Coast Guard is small enough you know someone wherever you go.

Tell the twenty-seven commanding officers that trip reports are to be no more than one page, twelve-point font, preferably with pictures. They are paid to command, not write novels. If they struggle, remind them the declaration of independence was one page.

Avoid the temptation to criticize any other career path. When grouping a bunch of tribes (ship, pilots, boarding officers, regulators, oil spill responders), you may be stereotyped by the insignia on your uniform. Elevate to the operational and strategic level. All are important. Learn what the others bring to the table. If it's boring, be thankful they're doing it and not you.

If you have dignitaries at a ceremony, do not use a bayonet to cut the cake.

Chapter 29

Trouble in Kauai: the Hawaii Superferry

THE DOG WAS JUST getting on its feet when we received word that trouble was brewing in Hawaii. A group of connected investors had procured a huge, high-speed "Superferry" with the goal of creating a run between Oahu, Maui and Kauai. Oahu and Maui were already heavily commercialized, and the people on those islands wanted the chance to visit the other less populated and less commercialized islands in the chain.

Kauai is a beautiful, pristine island. If you saw the movie *Jurassic Park III* when the boy is parasailing and gets stranded on the island. Well, that's Kauai minus the dinosaurs. There's only one main road, and the island probably looks  a lot like Hawaii from 200 years ago. A Coast Guard station on Kauai provided critical search and rescue coverage.

The islands in the Hawaiian chain had become militarized over the years, especially Oahu. It's the strategic midpoint in the Pacific. That's why we have a Navy fleet based there and why the Japanese attacked Pearl Harbor. During and following WWII, more and more military outfits poured in. The investor group in the Superferry project included several retired, influential, Navy admirals.

There had been a clear trend on all the Hawaiian islands that the native people were systematically losing the beautiful peaceful land to a growing tourist and military population. The bastion of Kauai would become the place to stand their ground.

While the Superferry's route was being challenged in court, they attempted to make their first run into Nawiliwili Harbor, Kauai.

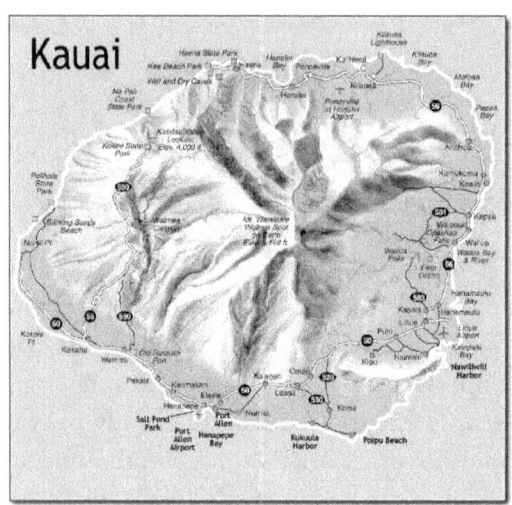

The Coast Guard's role was to facilitate free and safe commerce. After 9/11, it was common practice to escort vessels with large numbers of people, including ferries. The threat was a terrorist small boat attack, so the station deployed for the escort. In addition to the Superferry attempting to dock that day, approximately one hundred protesters entered the water, blocking the channel. They were in canoes, surf and paddleboards, and some just swam out to the harbor entrance. The massive ship was stalled offshore. When the station boat came out to intercede, the protesters wrapped their arms around the Coast Guard boat's propellor, not allowing them to engage without injuring them.

The stand-off was tense, but eventually, the ship departed without entering Kauai. It was a victory for the in-water protesters. But then they crossed the line and made death threats to the families of the Coast Guard station. That's when they called the DOG.

My job was to assess the situation on the ground and make sure we facilitated free and legal commerce while also protecting free speech and the protesters' safety. When I flew a reconnaissance mission out of Barbers Point, Oahu, in a helicopter, those on the ground in Kauai communicated with us via hand signals. They gave us the bird.

We landed and discussed with the local Coast Guard what could be done. The Fourteenth District Commander was Rear Admiral Sally Brice O'Hara, and her chief of staff was Captain Charlie Ray.[1] Their

---

[1] Both Sally Brice O'Hara and Charlie Ray served as Vice Commandant.

*In water protestors thwart the arrival of the Superferry by blocking channel
and wrapping arms around the Coast Guard's boat propellers.
(photo courtesy of DVIDS)*

area of responsibility included Sector Honolulu, which had the Hawaiian Islands and American Samoa, along with Sector Guam and the western Pacific. The sector commander for Honolulu was Captain Vince Atkins.[2] We had a great team in place, but this was so heated there were daily conference calls with the Commandant, Admiral Allen. This conflict was decades in the making, and we in the Coast Guard found ourselves in a precarious position.

The locals had a problem with commercialization and the military in general. The Coast Guard's reputation for search and rescue and environmental protection had thus far kept us in the neutral category, but we were now viewed as part of the enemy's team after the protest. Personally, if I were to take off the uniform, I could see being aligned with the protesters, so it's a difficult dance. But we're in the military, and the mission was to find a way to clear the protesters from the water. Bottomline, this had broader attention. External "professional protesters" like Green Peace were becoming involved and coaching the locals. There was a concern that if we let the protesters push the limits and win by taking these actions, there could likely be more in-water protests in the future, not just in Kauai. Could you imagine blocking supply ships bound for the Middle East to resupply our troops on the ground or blocking oil tankers in a narrow channel?

---

[2] Vince Atkins also made flag officer.

This was a precedent we needed to win. Our mission was to promote the free and safe use of the waterways, period.

In the book *The Chronicles of the Hawaii Superferry,* the authors do a good job of painting the broader picture of what was happening on the islands, but frankly, they lost my respect when they painted the Coast Guard as mindless robots following orders. They should have interviewed someone in the Coast Guard to get a balanced perspective. That said, I was in charge of the goon squad and was tasked to support Sector Honolulu with a specialized adaptive force package to get the ferry safely to the dock whenever the next run occurred, while simultaneously protecting the station, its families and the protesters.

When we arrived, there were all sorts of ideas on how to get the protesters clear from the channel. They ranged from chumming the waters to tactical plungers for surfboards to water cannons. All, of course would make great footage on national TV but would destroy any remaining remnants of our humanitarian image.

So, here's the problem. How do you clear approximately one hundred people—including children—some with spear guns, from the water for about an hour to allow the ship to come in and dock, unload and then depart? There was not much of an upside. Many bad things could happen. A protester could get injured or killed. Or, to avoid the protesters, the Superferry could run aground, creating a huge oil spill.

I pulled an all-nighter, writing up the tactics, techniques and procedures based on great ideas from the team and briefed RADM Brice-Ohara the next morning.

The plan utilized the sector commander's existing Captain of the Port authority to put in place a safety and security zone. That way, the commander controlled everything in the zone, which included the entire transit to the dock. There were healthy fines for people who tried to block the safe passage of ships in the zone, hopefully discouraging most from participating. But the zone would only be in place for an hour, for minimal disruption. So, knowing that there would still be some hard-core protesters, my specialized adaptive force package had to come up with a plan to physically move them out of the way. The couple of boats at the station were not going to cut it. We also knew that the protesters had figured out that wrapping arms around the props incapacitated the boats.

I remembered back to TACLET when we had gotten a black rubber boat for training called a combat rubber raiding craft or CRRC. You see SEALS and Marines running with these boats over their heads in all sorts of movies. I figured we needed six of those. Sure, they could be popped, but it would allow Coasties to come alongside swimmers and be less dangerous for everyone. Then we procured twenty-horsepower outboards (like I had on my Jon boat in Savannah) and cages that mounted around the propellers. That allowed us to counter their grabbing tactic.

Some of the protesters looked like Dwayne Johnson ("the Rock"). They were big Samoan-looking men. A petty officer came up with a great idea. With only two Coast Guard grabbers in the front and two boat operators in the back, you wouldn't want to put an agitated body builder between you. So, he put a hand cuff on the end of a rope. We'd come alongside, slap the cuff on a wrist, and back off while holding the rope. We would let the protester struggle like a marlin on the line. When he was good and tired, we would tow him to the beach and the local police department.

I briefed the plan and got a thumbs up to try it.

We identified three Maritime Safety and Security Teams to support: the local MSST Honolulu, MSST Los Angeles and MSST San Diego who had worked with on the Arellano-Felix case. The DOG logistics team also lined up an Air Force C-5 that flew our boats and teams to Honolulu.

The COs for all three MSSTs units happen to be female: Michelle Hoerster in Honolulu, Charlene Downey in Los Angeles and Rosemary Firestine in San Diego. When we made it to Hawaii, the Sector Response Chief Todd Wiemers[3] and his folks started calling us Charlie's Angels.,

Our plan was coming together. We found a secluded cove on Oahu and used a patrol boat as a simulated Superferry. We used Coast Guardsmen in the water to act as protesters, and we tested our plan with a stopwatch. We carefully filmed everything, and then I flew back to Washington, DC, to personally brief the Commandant. Admiral Allen gave us the go-ahead.

Upon my return, the Superferry had a test run from Oahu to Maui. We were ready, but there was literally only one protester who showed up. Maui was already commercialized. It was Kauai where the problem would occur. Just when we were ready to execute the plan, the courts

---

[3] Todd Wiemers eventually made RDML.

suspended using the ferry, sighting environmental concerns, namely high-speed ferries running through whale breeding grounds.

It may sound like going back and forth to Hawaii was a good time, but it was a pain climbing on the plane that last time. I was actually glad we didn't have to execute the plan, as it would look bad even if everything went as planned.

In the end, we had a good test of the fledgling DOG. We created a good plan for protesters in the water, which was improved upon and used later in Seattle, Portland and Houston. In those cities, the Coast Guard would not only have protesters in the water but also people hanging from bridges to block ship traffic. We had a plan for that too.

In closing and looking back on it, along with the St. Croix mission, one might think I was an anti-islander, on the side of commercialization and tourism. I'm not thrilled with those images, and frankly, I see why these people have a right to protect their lifestyle and property. The protesters crossed the line when they issued death threats and used violence against innocent people.

This was not the last time I saw the Hawaii Superferry.

Later in my career, the Coast Guard came up with the idea of establishing "First Amendment Zones" where people can present their message while remaining safe outside the main channel, and the waterway remains open. A joint meeting with a judge will ensure the protesters understand that they will face severe legal consequences if they do not follow the agreed-upon rules. This is good for the enforcers too.

Lessons Learned:

If a way of life is threatened, people become dangerously unpredictable.

South Park made a funny episode about this mission. Do an internet search for "South Park Coast Guard."

Sometimes simple solutions work best: rubber boats, cages around propellers and a handcuff on a rope.

Flying to Hawaii three times sounds like fun, but not really.

We can fit six, 25-foot, small boats on trailers in a C-5.

Consider creating a First Amendment Zone.

Chapter 30

Specialized Force Packages and FBI's HRT

ONE OF THE INITIAL missions for the DOG was teaming up with sister DHS agency the TSA for a Visible Intermodal Prevention and Response (VPIR) operation (pronounced like the snake). This was a tactic borrowed from the NYPD. The NYPD randomly sent SWAT teams to subways, airports and ferries throughout the city to disrupt any potential terrorist planning efforts. In a terrorist environment, you always want to be a harder target than the next guy.

Even before the DOG commissioning ceremony, my mission was to deploy to the Miami International Airport to determine if our canine explosive detection teams could be of value to airport security and our sister agency, the TSA. As you can imagine, there were serious legal questions regarding jurisdiction and tactics. My deployment partner was Brad Kieserman, an attorney who had been in the law enforcement office when I was at headquarters.

Helicopter delivery of our canine teams (photo courtesy of DVIDS)

When I first got to DC as the EA, I had friends in the building. One of my buddies, John Davis, was in the law enforcement shop, and he introduced me to this former boatswain mate chief who went to OCS and was now a lieutenant commander. He was—*how should I say it?* —a bit chubby, outgoing, but "perhaps the smartest guy in the Coast Guard." That last comment came from the Commandant. He had carte blanche permission to barge into the commandant's office at any time unannounced and by-pass the front office gauntlet of handlers and senior flags. He was in a special category I had not seen within the Coast Guard up to that point.

Almost every organization has its own version of Brad. Not only was he brilliant, but he also had an unrivaled work ethic and great people skills. He was the lawyer you would tell what you wanted to do, and he would find a way to get you to yes. He essentially wrote and maintained our Maritime Law Enforcement Manual, crafted numerous bilateral international agreements, and created the Maritime Operations Threat Response (MOTR) process—a "whole of government" system for responding to asymmetric threats on the water. They included responses to smuggling narcotics, migrants and weapons, and dealing with piracy like the *Maersk Alabama* hijacking. These events in international waters are extremely complicated. I had done the easy part of boarding and finding drugs. The hard part and, perhaps less sexy, was what you did with those you arrested, the drugs, the evidence package and the seized foreign ship. The list goes on, but it involves foreign governments, Departments of Justice, Homeland Security, Defense and State for starters.

At one of the many International War Games I attended, Admiral Allen said perhaps the greatest accomplishment during his career was creating the MOTR plan and Brad Kieserman deserved the lion's share of the credit. Brad wasn't even in the audience because, at the time, he had higher tasking as general counsel for FEMA. Brad was special, and the TSA needed help with the new program.

I traveled with Brad down to Miami. We started our mission with the DOG canine teams roaming through Miami's International airport. The Coast Guard canine handlers wore sidearms so there were jurisdiction things to consider like if there was an alert, if the dog bit someone, use of force and ability to detain, or even intercede in an altercation. Brad, like always, found a way to accomplish the mission.

I also led a DOG team on a VPIR mission to Maine to assist with security for the ferry system. Senator Susan Collins even came out for one of the ferry screenings. Forget the Coast Guard, TSA or CBP officers, the canine always gets all the love and attention. I deployed to similar missions in Seattle, Washington, and a nuclear powerplant inland

Canine team searches vehicles in Maine (photo courtesy of DVIDS)

in North Carolina.

I returned to Miami; this time with the DOG's Planning Chief Captain Merrie Austin. There we met with the FBI's HRT. The FBI had rented a large commercial ship, and we combined our specialized forces to stage a nighttime, full-scale, assault exercise launched from a 270-foot, medium-endurance cutter. The takedown offshore simulated a law enforcement response to a hijacked vessel or active shooter on a ship. FBI and Coast Guard boats and helicopters' communications and boarding skills were evaluated. The Coast Guard sector commander for Miami at the time was Karl Schultz, our future Commandant.

The FBI HRT was top-shelf for sure. Professional, fit and well-equipped. But ship boarding was new for them, especially responding in the offshore environment. They brought their flyaway package out of Quantico, a helicopter for delivery and small boats, but we also brought value as experienced boarding officers, armed helicopters, a ship with a big gun and, frankly, experience as maritime professionals. It was a good combination; for a large ship, you really need a large boarding team.

Boarding ships was the Coast Guard's forte and our full-time job. Even the

Admiral Karl Shultz

SEALs received refresher training from DOG units on boarding large commercial ships. SEALs do halo jumps and underwater ops and had been chasing Bin Laden in Afghanistan's caves. The FBI have been chasing terrorist for a decade and responding to the most violent crimes in the country. It's hard to keep proficiency in everything, but both organizations certainly earn their reputation as bad asses. Both the SEALs and FBI were extremely well funded and trained, but the Coast Guard and especially now the DOG brought a lot to the table. Coast Guard being both military and law enforcement allowed the DOG to serve as a coordinator and in some cases acronym translator between sailors and cops.

After we got these interagency VPIR missions going, the local forces started to do their own VPIR ops with local TSA, Immigration and Customs Enforcement (ICE) and CBP offices. I was eventually summoned to the Commandant's office with Admiral Atkin where the TSA administrator presented a plaque to commemorate the effort. I received it on behalf of the team, but it really should have gone to Brad. This was when Admiral Allen recapped my submerged car rescue. He actually made it sound more significant than it was!

Lessons Learned:

In any security operation or classroom, the canine will always get the most attention. Do not let them know it's just an explosive sniffing dog.

Make sure when you're doing a security drill at a nuclear power plant, the 911 emergency operators are clued in. Helicopters, checkpoints, and sirens tend to freak out the neighbors, per the feedback we received at the conclusion of the exercise.

Taking down a large offshore ship takes a big team, and seasickness can hinder the ablest team.

Chapter 31

Houston, We Have a Problem

FINALLY, AFTER FOUR YEARS of college and twenty-two years of service, there was a mission requiring calculus. To be clear, I was not asked to perform these calculations. The good folks at US Strategic Command (STRATCOM) and NASA did that.

STRATCOM is based at Offutt Air Force Base near Omaha, Nebraska, and is led by a four-star admiral or general. Their web page says the commander serves as the unified commander for all four branches of the military. That left me a bit irritated as we're the fifth branch, and we worked on their first operational mission as deployers. Space Force wasn't in the mix yet.

Their mission is strategic deterrence, nuclear operations, space operations, joint electronic spectrum operations, global strike, missile defense and other higher-level math missions.

Army Brigadier General Horne readies the troops for deployment anywhere between the Arctic and Antarctic circles.

This particular operation included chemistry, so right off the bat, I questioned whether they picked the right guy for the job. The point here is when you're grinding through those high school and college classes, there just might be a need to use it in the distant future. So, pay attention, students!

Your first question should be, "What the hell are we doing sending Coast Guard guys and gals on a mission involving spacecraft?" [1]

Judging by their recruiting commercials, this mission should clearly have been in the Air Force's lane. My older brother Johnny and father-in-law, Jerry, both Air Force veterans, assured me of that. But the answer is not as simple as you might think. If you read the National Contingency Plan and its annexes, you'll find the Coast Guard has a role in the recovery and response to environmental consequences of spacecraft. So how did I get sucked into the vortex and eventually serve as the J3 or operations officer for this mission?

It was a Friday afternoon in early January at the newly commissioned DOG. We had a little downtime over Christmas and New Year, but it was short-lived. We were back in the office again, operating rapidly, pushing deployable specialized force packages around the globe. It was getting toward quitting time on Friday afternoon when a call came in from Coast Guard Headquarters. I was ordered to immediately meet three admirals in the SCIF. The Sensitive Compartmented Information Facility was a space where you could discuss top secret issues. The DOG didn't have a space in our building in Ballston that met those requirements.

While it was Friday afternoon and DC traffic is *special* that time of day, I was intrigued by this mission's clandestine nature. I knew it had to be important because three admirals were extending their day to discuss it.

Due to the nature of this mission, I cannot reveal many details, but I was assigned to the National Reconnaissance Office (NRO) in Chantilly, Virginia, for the next couple of months. My mission was to serve as the Coast Guard liaison to Brigadier General Horne, the Army general in

---

[1] For the record, there have been two Coast Guard turned astronauts: Captain Daniel Burbank, who flew two shuttle missions, was one year ahead of me at the Academy, and while he was from Massachusetts, I wouldn't categorize him as a "masshole." Then there was Bruce Melnick, who retired as a commander. A 1972 Academy graduate, he was quite the heroic rescue pilot before he retired and went to NASA where he logged over 300 hours in space. The Coast Guard is very proud of these space pioneers!

charge. In a way, it sounded pretty exciting. I lived in Manassas, so it cut my commute in half. To a DC staff officer, you couldn't have asked for a better Christmas present. I thought this would finally allow me to see my family when the sun was out. I was wrong. I never worked harder in my entire career.

The next logical question was why me. Perhaps I was the most expendable, as the operations staff was filled to the brim with high performers. But probably the main reason was my top-secret SCI level clearance. At the time, very few had that level clearance, and with the DOG being new, it was good to link up with other agencies at the national level. After all, titles count, and Admirals Allen and Atkin were clever in naming our organization the *United States* Deployable Operations Group.

I couldn't tell folks at my normal busy job or my family what I was doing. It was kind of like being on a jury, unable to share any information. Only Admiral Atkin and my direct boss, the DOG Deputy Gary Rasicot, were privy to the initial details.

Despite a much-improved commute, this was no vacation. Fortunately, my deputy in DOG operations was my carpool buddy Keith Smith who covered the shop without missing a beat. In my extended absence, Merrie Austin, who was in charge of the DOG planning office, picked up my standby deploying status.

I have always considered Merrie a brilliant leader and friend. As a classmate of Astronaut Daniel Burbank and a bit of a chemistry nerd, she would have been a better choice for this mission, but she had her own opportunity to shine later during the Deepwater Horizon oil spill.[2]

It's been twelve years since the mission and much has since been declassified, and eventually, some portions were covered in the news. So, I'll stick to the unclassified portions and share a few personal observations. (I also had to clear this entire book with the intelligence community as well as DHS.)

When I arrived at the NRO, it was a major credentialing process just to get in the door. Rarely had I worked with classified material at this level, so this was a brave new world for me. I eventually met with the leadership team, which was in its infancy.

A top-secret US government satellite had been successfully launched into space. However, once in orbit, the onboard central processing unit

---

[2] Both Merrie and Keith commanded the Fifth District as admirals.

necessary to operate the thrusters to keep it in orbit stopped working properly. All attempts to control the satellite were lost.

The satellite's propulsion used a large hydrazine fuel tank designed to keep it in orbit for many years to come. But now the satellite that was travelling at 14,000 mph was coming down with a full tank of hazardous material (HAZMAT) and a classified payload. The scientists working on the problem determined the satellite would come down anywhere between the Arctic and Antarctic circles, and the payload and dangerous fuel in a titanium tank would survive reentry all the way to the surface. In the unlikely chance it were to hit a city, there could be a significant loss of life.

The plan was to form a multi-agency joint task force named Operation Burnt Frost. This particular force of fifteen agencies led by General Horne would respond anywhere in the world (between the Arctic and Antarctic circles) to this threat, recover the payload and contain and/or remediate the hydrazine fuel. Piece of cake!

Using level-alpha personal protective equipment, our personnel would build a coffin-like containment system around the tank. We would also remediate any HAZMAT that leaked out. Essentially, it's our garbage, and we'll be responsible for cleaning it up if it lands in a neighbor's yard.

As a liaison, and organizationally the supervisor of three pollution response strike teams, I was able to reach back to bring in Coast Guard experts and procure specific equipment and chemicals for this classified mission.

As you can imagine, packing for this trip was difficult. Suntan lotion or parkas? A tourist camera or an M16? Communication equipment was different for different hemispheres. We might have to go to Afghanistan in the winter or Brazil in the summer. Let's not forget the shots—there were a lot of those.

The NRO was busy assembling 150 experts which included all military services, EPA, and a host of intelligence folks. We even had a NASA space shuttle pilot and a National Parks superintendent. My experience to date had absolutely nothing to do with space. I had been chief of the International Training Team and travelled extensively. I had just come from being the commanding officer of PACTACLET and was a former group commander, so I had some applicable interagency experience on the surface of the planet. What I lacked was expertise in pollution response, specifically what to do with a tank of hydrazine.

In full disclosure, my second worse class behind calculus was chemistry. Fortunately, upon the creation of the DOG, the three National Strike Teams worked for Captain Rod Walker who organizationally worked for me and of course RDML Atkin. In building the Operation Burnt Frost team, we got fifteen talented strike team folks to support the task force. They included two noteworthy officers who later rose to the rank of admiral and captain. Rear Admiral (Select) Mike Day[3] and Captain Monica Rochester[4] were assigned to strike teams. LT Mike Weaver[5] was also in the mix.

Fortunately, I had just completed Naval War College at night, which proved helpful, and I served as a translator between DOD jargon and the civilians on the task force. That, coupled with a bit of energy and urgency, I drafted up an alert on what to do if the satellite landed "in your country." That document (you should always spell check and ensure classification hurdles are jumped through properly) was published at the United Nations.

JOINT TASK FORCE BURNT FROST
McGUIRE AIR FORCE BASE, NJ
17 FEBRUARY 2008

Mark Ogle is the middle holding the clipboard, next to the general
(certificate deployers received)

---

[3] Rear Admiral (Select) Mike Day was a bit of a folk hero during 9/11 orchestrating the Manhattan boat lift and would later command Sector New York.

[4] Captain Monica Rochester became the commander of Sector Los Angeles.

[5] Mike Weaver later helped me get a retirement job on his staff.

I linked up with a very bright Army command sergeant major, and we took it upon ourselves to start drafting up the actual operations plan (OPLAN) of how this force would deploy and what we would do when we landed at the crash site. Our urgency impressed General Home. Based on this, General Horne wanted to show the diversity of the joint task force and asked Admiral Atkin if I could serve as the J3 or chief of operations for the mission. Eager to get the new US DOG's attention, he agreed.

In one of those first all-hands, General Horne made the announcement. "This is Captain Ogle of the Coast Guard who will be our operations officer. He doesn't know anything about space, so if he can do it, we all can."

If you 're wondering, the deputy was an Air Force colonel.

With that glowing endorsement, we worked around the clock, refining the OPLAN, engaging all the COCOMs, running exercises, getting supplies and even preparing briefs for President Bush. COCOMs are militarily responsible for the seven geographic areas of the world and are each led by a 4-star admiral or general. We also worked with functional COCOMs like STRATCOM, SOCOM, TRANSCOM, and now they even have CYBERCOM and SPACECOM.

Some of these folks have since changed their names, but we worked with them all. For example, if the satellite went down in South America, our force would fall into USSOUTHCOM's area of responsibility. We'd get there by lining

Briefing actions to retired Admiral Harvey Johnson, FEMA, should the satellite debris land in the US

up aircraft from USTRANSCOM. The operations officer of a COCOM is normally a one to three-star admiral or general, so being the boot captain J3 Coastie of a unique interagency gaggle operating at the top-secret level was challenging. Planning for and supporting this inbound task force would require a lot of effort on their part for an operation that "probability" indicates wouldn't happen in their area of responsibility.

At one point during the process, General Horne and I briefed the chief of operations for FEMA to prepare in case of a domestic impact. At the head of the table sat a familiar face. It was now retired VADM Harvey Johnson. As I passed him, I told him I was back carrying the clipboard again. He was actually excited to see me. After the brief, we talked about getting General Horne out on a Coast Guard speed boat.

While the task force drafted TPFIDs, which is DoD slang for packing orders, we accumulated final equipment and packed two dedicated large cargo planes at McGuire Air Force Base in New Jersey. Simultaneously, a separate group within STRATCOM worked on the calculus to shoot down the satellite in low orbit.

The missile would be flying at 3,000 mph and had to hit the small satellite flying at 14,000 mph. This reminded me of one of those multiple-choice SAT problems!

The USS Lake Erie fired the missile which destroyed the malfunctioning government satellite in low orbit over the Pacific. (photo courtesy of DVIDS)

If you've seen the movie *Gravity* starring Sandra Bullock, an explosion in upper orbit creates a huge debris field which takes out the space station. That movie was loosely based on a real event where the Chinese fired a missile into space to destroy one of their satellites in high orbit.

The United States publicly criticized China due to the hazards of the debris to satellite constellations and future space launches. There was also a second concern; China had a proven capability to shoot down satellites.

The USSTRACOM mission was to upgrade the three heat-seeking missiles on the USS *Lake Erie* in the western Pacific to be capable of entering space and destroying the wayward satellite in low orbit before it tumbled in re-entry. Shooting it over the Pacific would minimize the chance the satellite would come down on land or in a populated area.

I lost about fifteen pounds from nonstop work, missing meals and ass-chunk removal while preparing the mission. The satellite was approaching the atmosphere; it was time to take the missile shot. A big take away: don't shoot missiles into space without picking up the red phone telling other countries like Russia, China, and North Korea. They need to know this isn't a preemptive attack to start World War III. That said, you also don't want to tell them too early and give them the ability to set up observation points.

President Bush approved the plan, knowing full well success was far from guaranteed. This kind of operation had never been done before from the surface by the United States.

Our only break for the month had been to watch the Superbowl, and I fell asleep during it. The task force had a great team, but I was ready for this mission to end. We had cots in the terminal, and we had the planes packed. The only way we weren't flying somewhere to chase parts was if we had a direct hit. In the command center set up at McGuire Air Force Base, the mood was tense. They mounted a camera on the missile so we could see the real-time launch. Everyone in the command center knew that if the missile strikes were unsuccessful, we were boarding planes.

The first missile was perfectly guided in for a direct hit! That would have earned an A+ in calculus. The room erupted in relief. The only one mildly discouraged was our leader, the great General Home. He was responsible for that portion of the mission too, but he was a warrior at heart, and he was ready to deploy. He had moved heaven and earth to get this team and equipment together, and he wanted to see them in action.

We succeeded, and I would argue this mission contributed to the need for Space Force!

Here is an abbreviated unclassified excerpt from Wikipedia which covers the fallout.

> Operation Burnt Frost was a military operation to intercept and destroy a non-functioning U.S. National Reconnaissance Office (NRO) satellite named USA-193. The launch occurred on 20 February 2008 at approximately 10:26 p.m. EST from the USS *Lake Erie*, which used a Standard Missile-3 (SM-3) to shoot down the satellite. Only a few minutes after launch, the SM-3 intercepted its target and successfully completed its mission, by neutralizing the potential dangers the errant satellite originally imposed. While the threat was mitigated, Operation Burnt Frost has received much scrutiny from other countries, mainly China and Russia. Following the operation, China and Russia criticized the U.S. operation. The destruction of USA-193 came just weeks after Russia had drafted a new treaty to ban space weapons which were backed by China at an UN-sponsored forum. This treaty would ban the use of weapons against satellites or other spacecraft. This prompted the Russians to accuse the United States of using the hydrazine gas as a cover-up to test an antisatellite (ASAT) weapon. They claimed that several countries' satellites which used toxic fuel have crashed into the Earth in the past but never warranted such "extraordinary measures." Furthering this notion, others have speculated that the toxic gas would have likely not survived re-entry regardless and, even if it had, that the risk would be extremely small. These speculations have led many to believe that Operation Burnt Frost was in response to China's ASAT test on 11 January 2007, and the fear this would begin another "space race."
>
> However, unnamed U.S. officials continued to deny that the shooting down of the satellite was in response to China's ASAT test one year prior, or that they were trying to protect classified satellite technology. To promote transparency, the U.S. delegation to the Committee on the Peaceful Uses of Outer Space stated that after the operation concluded, the special modifications made to the two remaining technical

missiles and three naval vessels were removed and that the United States "has no plans to adopt any technology from this extraordinary effort for use on any current or planned weapon system." U.S. officials pointed out that the U.S. had no reason to prove that it could shoot down a satellite, as the U.S. had already publicly done so in the 1980s. Another key difference pointed out by General Cartwright was that this intercept happened at a much lower altitude, whereas China's ASAT weapon destroyed a target at a much higher altitude, which resulted in the creation of debris which continues to pose a potential hazard to other spacecraft. Finally, U.S. officials again affirmed that the mission intended to preserve human life.

At one of the final all-hands where all five services were present along our civilian deployers, General Horne wanted to hear each services' war cry. Army (hoorah), Navy (hooyah), Air Force (hooah), Marines (oorah), and then it was our turn. I think it was Mike Day who responded.

"Sir, ah, we don't really have one of those."

I think the general was disappointed. Something for the Coast Guard to work on.

Lessons Learned:

I guess Space Force will handle this type of mission in the future.

"Wire brushing" is what Army colonels call an unpleasant session with a general.

A titanium hydrazine tank that powers a satellite will make it to earth's surface.

When using modified heat-seeking missiles to shoot down a satellite in orbit, its best to let it heat up over the earth's largest ocean, the Pacific, and hit it before it starts tumbling in re-entry.

Last-minute packing for anywhere between the Arctic and Antarctic circles is challenging.

Bringing a "whole of government team" together to tackle a never-before-seen problem was pretty cool

Chapter 32

Fourth Command

SECTOR HAMPTON ROADS, COVERING ALL OF VIRGINIA AND COASTAL MARYLAND

TWO THINGS INDICATED I had finally reached the big league:

1. For the first time, this former executive assistant has his own administrative assistant.

2. While briefing an admiral, my cell phone rang, and I had to take it. It was the governor asking me when I was going to let the Navy's Second Fleet back into port.

The sector command structure was new for the Coast Guard. During 9/11, Virginia, for example, had multiple Coast Guard commands overlapping in the same area. Captain Larry Brooks was the Captain of the Port and commanded the Marine Safety Office which controlled and inspected commercial shipping, anchorages and credentialing of mariners and responded to pollution events. Larry was extremely competent and effective but didn't fit the stereotype of a modern politically correct Coast Guard captain.

At the time, he was a chain-smoking New Yorker who didn't have much of a filter. He knew his stuff and had no problem calling out senior admirals—Navy or Coast Guard—if they were wrong. His authority extended over three group commands: Eastern Shore, Hampton Roads and Cape Hatteras. Being from completely different worlds, little did I realize that he and I would become great teaching buddies.[1]

---

[1] He retired out of this command tour and designed a course for sector commanders and sector department heads.

While Larry's authorities were vast, he didn't have direct control over Coast Guard boats, ships or aircraft. Those assets fell under the groups like Eastern Shore that I commanded. Unity of command is important, so there had been a prototype consolidation of commands in New York and Baltimore called activities. Both activities had performed well during the 9/11 attacks and served as a one-stop point of contact for all Coast Guard services. The commander had both the authority and tools to do all missions. So, there was a consolidation plan for the rest of the nation, something I had worked on as an executive assistant. The groups and Marine Safety Offices went away, and thirty-seven sectors were born.

The chief architects for these sectors were three captains: Larry Brooks in Virginia, Roger Peoples in Baltimore and Dean Lee from my old carpool. After the 9/11 attacks, the Coast Guard rolled out the consolidated units in 2005. When I received orders to command Sector Hampton Roads in 2009, I knew roughly half the job. The rest I learned from Larry over the next seven days.

Normally, that learning would take place over a twenty-year career. As great a teacher as he was, one of Larry's greatest career achievements was hiring Mrs. Mary Stevens at his unit in Hampton Roads. She became the sector's administrative assistant. Media and the Hollywood crowd often portray government workers as lazy, not-that-bright, government bureaucrats. I'm sure there are some out there, but most I've met were just the opposite. Mary Stevens was one of the best.

Mary had been at the top of her class and was a wonderful lady with common sense and even better people skills. She was perhaps ten to fifteen years older than I was. Side by side, she ran circles around me. I loved to take walks and get her opinion on things. I told her my goal was to take this large organization of diverse specialists and get them working toward a common goal.

Mary recounted President Kennedy's challenge almost a decade earlier.

Mary told me a story about a reporter looking for a story. The reporter entered the Cape Canaveral facility late in the evening. He wanted to do an interview, but the place was a ghost town. The only person he could find was an older man mopping up the bathroom.

He stopped and asked, "So, what is your job here at NASA?"

The old man answered, "It's to put a man on the moon."

So how do we create that culture like that at a diverse sector? Not easy but it starts at the top, as Mary would say, so you best stop walking and get back to work.

I did have one complaint about Mary: she was from Massachusetts. Now she wasn't a *masshole*, mind you, but she loved the New England Patriots. It was an incurable sickness. During our TACLET tour, my son and I became avid San Diego Charger fans to the point where we had flags in the yard, stickers on the cars and a full-size helmet on my desk. That poor helmet. One time after a very disappointing Sunday night performance, Mary taped Patriots' decals over the lightning bolts. Another time, during a white elephant gift exchange, the helmet emerged out of a box, and everyone looked at me, camera phones focused, waiting for my reaction.

That helmet helped humanize me and get to know the 2,000 active duty, reserve, civilian and auxiliary personnel from all over the country. They were now part of Sector Hampton Roads. I didn't want to be that guy who could only discuss work—although there was plenty of that. I wanted to connect on a different level if possible. Mary and this helmet gave me an idea, an opportunity, and I developed an underlying strategy to engage people.

The Sector Wardroom and families went camping and tubing down the Rappahannock River.

I had an analogy I used to describe the teamwork and roles of this new complicated organization. I compared a sector to an NFL team.

You see, the Coast Guard was like the National Football League (NFL); the Commandant was like the NFL commissioner and set policy. Then you had two areas: Pacific and Atlantic, like the American Football Conference and the National Football Conference. The NFL had divisions with four teams in them. We had districts that had about the same number of sectors. There were thirty-seven sectors and thirty-two NFL teams spread across the country, representing certain populations. That was just the setup.

Once you looked at the sector, the command was like the coaching staff.

Remembering back to Naval War College and advice I received from admirals at headquarters, it was time to elevate out of the tactical and even operational levels. It was time to start being strategic. The sector departments aligned with the NFL team's coordinators: offense, defense and logistics. Intelligence was like those who scout the other teams. Planning wrote the playbook. The command center was like the quarterback or linebacker. Reservists were back-up players. The sixteen units were the players.

During in-briefs of new personnel, I always came out from behind the desk and sat on a couch in the meeting area. The helmet was on the table. After hearing about their move and their career goals, I explained their role within this broader team concept and what my expectations were. Our goal, of course, was to keep our fan base (the public we serve) safe and protected. If we found ourselves in the Superbowl, which would be like a hurricane hitting the port, we would prevail. I very much saw my role as a coach.

The coach gives out a game ball following success. Maybe that game ball was a timely medal.

The players should be the face of the media when things go well; the coach is the face of the organization when things don't.

The coach also needed to keep distracting drama that happens off the field to a minimum.

I got a sense of how I was doing when the team wore their jerseys during off-hours, put Coast Guard stickers on their cars, and sported unit t-shirts.

One thing I had never liked was giving speeches. Someone took a poll of the greatest human fears; public speaking was number one, followed by death. Frankly, I'm less scared during a midnight drug boarding than standing in front of a big audience, yet now it was part of the job. One of the best speakers I had ever seen was Dean Lee. He captivated audiences by telling stories to get his message across instead of remembering boring lists, dates and 200-year-old quotes.

It was time to take command for the fourth time, and RADM Lee's change of command *speech word* was back again. At the Deployable Operations Group, we had a committee of current commanding officers whose express mission was coming up with a particular word for incoming commanders to use in their change of command speech. It was a lot of fun when you weren't the one taking command. Now the shoe was on the other foot. The word given to me for my speech at Sector Hampton Roads was "Sham-Wow."

I was relieving a genuinely great officer named Pat Trapp. He was well-liked and quite the character and caterer. With a crowd of a couple of hundred guests, the relieved commander personally foots the bill for lunch for all the guests. It was like being the father of the bride. He served shrimp and I think I even saw lobster tails. He was setting a high bar across the board! He was friendly and energetic, and that's how I described him in my speech. I said my first impression of Pat was a cross between a game show host and a sham-wow salesman. I got a laugh because it wasn't far off; Pat had a friendly way of selling a plan. He was a good sport even though he didn't know about the speech word at the time.[2]

It had been an exceedingly long week to relieve Pat officially of his responsibilities. The command included 550 active-duty personnel, 180 reservists and 1,300 auxiliarists broken into five divisions. There were seven stations, six cutters, three aids to navigation teams, which all had their own officer in charge or commanding officer. With the exception of Station Indian River, Delaware, I once again had command responsibility for the Eastern Shore.

That week, I met with every agency in Hampton Roads. Nye and the kids, my mom and her boyfriend Jim, and Nye's folks all came in for the ceremony and departed shortly after. Both kids were still in school in northern Virginia, so Nye did the solo parent thing for a couple of months. I had to relieve early in April because Pat was shifting into his new role which was vacant due to a retirement.

---

[2] Pat Trapp became the chief of staff of the Fifth District and my immediate supervisor.

The first night, I was living in a campground in our RV and having some electrical issues. I was so tired, I decided to worry about that in the morning and quickly fell asleep. However, my new cell phone rang at midnight. After being in command for all of twelve hours, I was being told there was a good chance the port would be destroyed that night.

Sector Deputy Jeff Novotny and the command center watchstander were on the line.

"Captain, you may want to sit down for this one," Jeff said.

I didn't tell him I was laying down already. It sounded as if they had done these many times before.

"Sir, a thousand-foot ship inbound to the Chesapeake Bay Bridge-Tunnel is a mile out."

All good so far.

"The pilot was on the ship."

Check.

"They have lost all power."

Bummer, like my RV.

"The current is still flooding in for the next two hours."

There was no way to turn or stop a 1000-foot ship in a mile.

"The pilot *thinks* he has enough momentum to get over the bridge-tunnel complex at the mouth of the Chesapeake Bay."

For the record, the ship was longer than three football fields. It would be super bad if it hit the Chesapeake Bay Bridge-Tunnel—originally one of the seven manmade wonders of the modem world—and cut off vehicular traffic to the DELMARVA Peninsula, blocking all maritime traffic to the Chesapeake Bay. That would include traffic bound for Norfolk, DC and Baltimore and block in the world's largest naval base. That would be super bad!

We shut down the vehicle traffic on the bridge, and sure enough, the pilot was correct; the powerless ship just made it over the tunnel before losing momentum and steerageway. The crew dropped the anchor. The issue now was the flooding current had spun the ship around, so the stern was facing away from the tunnel. My next question, of course, was how much chain was out and have we calculated the swing circle for when the tide changes.

We had only two hours, and it looked like, if the ship swung, it could possibly hit the bridge. It was going to be close, *if* the anchor even held. The sector team, along with the partners in the port which included the ship's captain; pilots and shipping agents, sortied tugs to retrieve the powerless ship and get her safely moored to a pier by two in the morning. The port I had been so overwhelmed by as an ensign twenty years earlier lived up to its billing. The port never slept, and that foreshadowed my next three years.

As I settled in that first week, I received a request from a very nice reserve Chief Warrant Officer Sharon Doggett, who was retiring that next weekend. She asked if I could preside over her retirement ceremony, and my default answer is always I would be honored, even though I had never met her. I had officiated several retirements at this point in my career, and usually, there were twenty to thirty people. It involved some remarks on my part, but the show is for the retiree. After their words, a presentation of gifts, awards and a shadow box with all their medals followed. The military does a nice job of sending off retirees.

Since we had never met in person, I asked her to sit with me for a while so I could get to know her and find out some details of her career. She was currently an Air Force civilian working at Langley but had been in the Coast Guard Reserve for thirty years. She was never married and had no children. When I do retirement remarks, since family generally makes up half the audience, I try to involve stories about them. This was going to be a bit trickier. I also like to go back in time and look at what the top movie or songs were for the year the person came in. I could tell right off the bat that Sharon was a kind, warm-hearted, professional woman. I wanted to do a good job, but it was certainly going to be a challenge coming up with content.

I drilled into hobbies and her career in the Air Force. I was not 100 percent confident, but I had a rough game plan. I figured there would be a handful of coworkers and probably her elderly parents. General Paxton, who was the Marine Corps general in San Diego had a saying: if you must write out and read a speech, you're a weak leader.

So, I did my best to commit my thoughts and details to memory. I had an early experience where I wrote out every word. When I was delivering the speech, I lost my place, and it was a disaster. Never again. With few exceptions, from then on, I tried to rehearse and commit to memory the content.

Sharon had booked the Botanical Gardens in Norfolk for the event on a Saturday morning. Unless people really like the retiree, holding a ceremony on Saturday morning usually thins out the crowd. In practice,

I always arrive early, and additionally beach traffic can be a nightmare.

When I went into the venue, my heart sank through my chest, through my knees and out the bottom of my feet. The venue was packed to the gills. I estimated at least 500 people and a formal lunch. We had a color guard with representatives from every service; we even had a Tuskegee Airman. I quickly went back to my car and started scribbling some notes on a sticky pad. As I was trying to revive from my panic attack, I realized something. She had never had her own baby shower or a wedding of her own. This was a big-time life's moment for her. I could not screw this up.

On stage, I was still churning up ideas. When it was my turn to take the podium, I took a big risk. I'd never had an audience of this size. I asked everyone in the audience, who physically could, to please stand up. That even included the WWII Tuskegee Airman. Those guys are pretty tough.

Then I told the crowd, "If you'd never had a part-time job in your life, please sit down."

A bunch of younger kids sat down and perhaps a few others.

Then I asked those who had a part-time job, but for less than five years, to sit down. Then ten. Then twenty. The standing crowd was starting to get very thin. When I got to twenty-five, we just had a handful. When I got to thirty, Sharon was the only one left standing.

I said, "That is why we are here today. To honor our friend and colleague, Sharon."

I went on to say, "Everyone here wants the same thing out of life, and you can boil it down to one word. Do you know what that is?"

There were quiet discussions but no answers.

I said, "It's to be happy. There are four components to our happiness, all starting with the letter L. To live, love, learn and leave a legacy, and Sharon, you have excelled in each category."

I left there relieved that my gamble had paid off. The ceremony even made it into the paper, and on the anniversary of her retirement, Sharon, being the class act that she was, called to thank me once again.

Another stressful moment for me was morale day at Harbor Park where the Tides baseball team was playing. I had been invited to throw out the first pitch. We had a huge contingent from the command in the stands along with my family. At the time, my twelve-year-old son Taylor was very much into all things sports and had practiced with me for the better part of the week.

But I wasn't the only one on the mound at Harbor Park that day. A young, special-needs man was scheduled to throw out the first pitch, and then I would be announced as the captain of the port. My partner on the mound was very friendly and surprisingly composed—living in the moment.

When he stepped up, something flipped in this kid, and he got really serious, in the zone. He threw a beautiful strike. The crowd cheered. I patted him on the back and said something like, thanks for making this harder on me. It was my turn. The key was not to try to burn it in there, just get it over the plate, and I did. I think it was also a strike, maybe just a little outside. I was relieved to be off that mound.

The next big event was a Veterans Day speech at my eight-year-old daughter Storm's Westside Elementary School. The principal, who was one of the parents on Storm's swim team, approached me to do the speech. I began thinking, it's just a bunch of kids. If I got through Sharon's retirement and the baseball game, I could do this. I just needed to speak at a young child's level, which I did every night.

Veteran's Day speech at my daughter's elementary school with vets who saw action back to WWII

So, I headed to the cafeteria the morning of Veterans Day where there were probably fifty veterans from all services, including veterans of all the wars from WWII to Iraqi Freedom.

Nerves started to kick in just a little when they started to line us up. I was in the front with the WWII veteran. The event was to be held in the school's gymnasium, but we had to go through a maze of hallways to get there. As we entered the first hallway, kids were lined shoulder to shoulder on both sides, waving American flags and cheering and shaking our hands. The veterans were emotionally moved; some were even crying. Perhaps I was too, but mostly because I had to give another speech in a few minutes.

The cheering got louder and louder, and when we entered the gymnasium, the packed house of screaming kids was deafening. Now

Mark Ogle

it was time for me to speak, so I went back to the audience engagement plan and had all the military dependent kids stand and be recognized and charged the rest of the seated classmates to look after them as their mothers and fathers would be away defending us all.

Then I went through war by war, recognizing those vets on the basketball court with me and used the sheep, sheep dog, and wolf story. That story describes people of the world in three categories. Ninety percent of the people are hard-working, good people who would never harm their neighbors. Those people are considered the sheep. Five percent of the people are considered dangerous, bad people who, given the chance, would attack and eat the sheep. These are the criminals and evil people of the world characterized by the wolf. The final 5 percent of people are characterized as sheep dogs. To the sheep, the sheep dog looks a lot like a wolf. They have the same shape, have fangs and growl. But the difference is they would never hurt the sheep. They keep the flock safe from the wolves. The sheep are still uneasy with the presence and appearance of the sheep dog, except when the wolf is nearby. That's when the sheep dogs are most appreciated. Those in the military and police are our society's sheep dogs.

In the end, it was an incredibly good experience, and most importantly, I didn't embarrass my daughter—too much.

Secretary of Homeland Security Janet Napolitano came to town for a special forces' demonstration. The Maritime Security Response Team (MSRT) was, at the time, our sole counterterrorism unit and was just down the road in Chesapeake. I got underway with the secretary on a buoy-tending ship that was a simulated hijacked vessel. The MSRT would demonstrate a coordinated boat and helicopter assault. Not only was I the captain of the port now, I knew this stuff from my previous jobs.

CAPT Mark Ogle meeting DHS Secretary Janet Napolitano

After the guys did a flawless takedown, we adjourned to the cutter's mess deck with the secretary. That's when she asked me point blank why the Coast Guard needed this capability when we have the SEALs and the FBI's Hostage Rescue Team. A wrong answer to a question like that could have negative career consequences.

A good stall tactic is, "Ma'am, that's a great question."

I told her I had done operations with the FBI's HRT and the SEALS. I also told her they were truly outstanding teams, but the SEALS were getting overwhelmed with inland overseas missions.

In fact, the TACLET teams and MSRT were now taking on piracy and waterborne threats in the Middle East. The SEALs had a precious, finite number of operators and were often needed elsewhere. The FBI's HRT based in Quantico was topnotch, but they were not specifically maritime focused. Their helicopter and their maritime personal protective equipment were somewhat limited. The final point I made described the joint exercise we did in Miami. If you want to take over a large ship, you need an exceptionally large boarding team. We are all human, and even the best operators like the FBI's HRT can be incapacitated by seasickness. The Coast Guard team understood shipboard living, systems, and what it took to board and operate ships. It only made sense for this capability to reside in *her* department. I dropped the mike and preserved my career.

Things were going well. Since we had a huge Coast Guard presence in Hampton Roads, there was an initiative in the works, driven by the locally based admirals and their staff and my counterpart Captain Fred Sommer who was the base commander, to recognize the city of Portsmouth formally as a designated Coast Guard city. There was a great deal of advertising going on, but it was summer, and I was super busy with search and rescue cases and happy to let others take the lead.

A morning talk show with Portsmouth's city manager and Fred was scheduled to discuss the upcoming events. The night before the "live" show, I received an urgent call from Fred saying he had a massive zit on his forehead and really needed me to cover the show for him.

Having had inopportune outbreaks in the past myself, I understood.

He said, "Your role is really just window dressing as the town manager has all the details."

Based on that, and Fred having been a great partner, I said I would cover it. After all, I was acne-free at the time. So, as always, I rolled into the studio early at 0700. The "live show" was not scheduled to start until 0800. I briefly met the host and hung out. Everything was cool, but I wanted to speak with the city manager before we started to see how I could best support his message. It was now 0730 and no manager. Then 0745 clicked past on the clock and still no manager. At 0755, a call came into the set saying he wouldn't be able to make it.

I had a flashback to Sharon's retirement. I really had only a couple of options. Number one, I could fake a seizure, or number two, I could filibuster with stuff I knew like Dr. Carafano 's war college class. I didn't have any facts about the events, so when it started, I essentially wrestled the microphone away from the host and filibustered for fifteen minutes, talking about recent cases and how the city of Portsmouth had been so good to the Coast Guard. I had survived another close call, but I doubt the host wanted me back.

Later that week when I was leaving a Coast Guard city planning event, a dark-colored sedan pulled up to the curb in front of me. The window came down, and sure enough, it was RADM Dean Lee, the Fifth District Commander. He was in the passenger seat and had a young LT driving him. RADM Lee is probably one of the best leaders and speakers in the Coast Guard. He has a booming deep voice, and although staunchly religious, he knew how to let his hair down—figuratively, because he was bald. I had fond memories of my current boss watching those Jackass videos on the way home in the carpool. He was also in incredible physical shape for an older guy.

On this particular occasion, out the window, he said, "Mark, we should do that Coast Guard City one-hundred-kilometer bike race, you and me."

I thought of my Huffy covered in dust and then looked at the driver who was sinking in his seat. I pointed at the driver with the rope on his shoulder. "What's wrong with that guy?"

RADM Lee chuckled; he knew I'd be there.

How hard could it be? I had to do six, twenty-five miles and one, fifty-mile bike trips for the cycling merit badge thirty years ago. Now I was thirty years wiser.

On the day of the race, I was there early and lined up looking for RADM Lee. VADM Rob Parker, the current Atlantic Area commander, spotted me and came over. "Hey, did you hear what happened to your boss last night? He got mauled by a Portsmouth police dog. Have a good race—flat, fast and fun," which had been the race motto.

I didn't get a chance to find out Dean's status, but I did have the next five hours to wonder what exactly he was doing when he got mauled. I linked up with a station petty officer who kept me going. The race fulfilled one-third of its promise; it was relatively flat.

Later, I found out Dean Lee had been riding his bike back from work when he turned the corner near a small park. The police dog was off the leash doing his business and was startled. Dean got pulled off the bike before the canine handler could react. The apologetic police officer had many children, and I'm sure he was thinking he'd be sued. That was not Dean's style.

I did ask the Portsmouth mayor privately at an event why his dog tried to eat my admiral.

In dealings with other agencies, I had hit it off with the state's Secretary of Transportation Sean Connaughton, a former Coastie and urinalysis coordinator as well as the 2006-2009 administrator of the United States Maritime Administration.

My other state partner was the Homeland Security Secretary Terrie Suit. She made sure Nye and I were at least introduced to Governor McDonald at a dinner he hosted for a large crowd.

The governor was very gracious and also a lieutenant colonel in the reserve, so he had a special interest in supporting the military in Virginia. Along with the other Virginia Army, Navy and Air Force base commanders, I served on the governor's military advisory council, which met periodically. This was a cool part of the job as sector commander. At one of those council meetings, we talked about Boy Scouts. Over half of the colonels and captains had been Eagle Scouts.

Mark Ogle as Captain of the Port of Virginia briefing International Delegation.

Lessons Learned:

If you close the port with a 240 million dollar-a-day economic impact, make sure you have a lot of partners on the conference call supporting the decision.

Buy a helmet for your desk and lock your office at night.

Never underestimate the size of the audience.

Never take speaking engagements lightly.

If you're throwing out the first pitch in front of a crowded stadium, for God's sake, practice!

If the secretary of Homeland Security asks why the Coast Guard needs its own counterterror unit, be ready with a good answer.

If you are an executive invited to a meeting and told you don't have to say anything, always be prepared to say something.

If an admiral suggests a 100-kilometer bike ride, consider upgrading from the Huffy.

When you get to have dinner in the governor's mansion in Richmond, you win brownie points with your wife.

Chapter 33

Storm Response

SECTOR COMMANDERS HAVE BROAD authority and responsibilities over a geographic region. The thirty-seven sector commands in the country include coverage inland and all the US territories.

Sector commanders wear five different hats. The captain of the port holds the broadest authorities with the goal of safety, security and environmental protection. The jurisdiction extends twelve miles offshore and can control ship movements, operations at waterfront facilities and even close the port down.

I used the doctor's hat to signify the goal of running a healthy port and solving challenges like shoaling, sinking ships, anchorages and special events.

The second hat worn by the Sector Commander is the search and rescue mission coordinator. Extending offshore to 200 miles and potentially in shore during flood events, the primary role is to save lives and property.

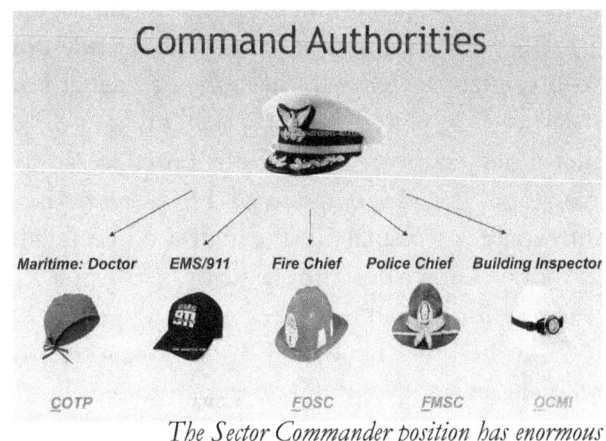

The Sector Commander position has enormous authority and responsibility

The federal on-scene coordinator authority covers response to oil spills and HAZMAT releases, including consequence management of weapons of mass destruction. This authority covers the coastal zone and extends offshore for 200 miles.

The federal maritime security coordinator is like a maritime police chief and focuses on prevention and response to criminal and terrorist threats.

The final hat is the officer in charge of marine inspections. This focuses on ensuring U.S flag vessels and credential mariners are following regulations. It also covers investigations into accidents.

While that sounds impressive, there are generally a handful of movers and shakers in a seaport. These are not the Coast Guard or Army Corps leaders who come and go every few years. These are the folks who have dedicated their entire careers to that one area. Two such men in Virginia were Bill Cofer and Bill Burket.

Captain Bill Cofer was the president of the Virginia Pilots Association and the Virginia Maritime Association, an organization of 550 port-based businesses. Bill Burket was a Virginia Beach Fire Chief who created and served as the director of the Maritime Incident Response Team, which was part of the Port Authority.

Cofer was a polished guy who looked like he could play a governor's role in any movie. He was an ideal leader of a white-collar, executive-level team. If something was happening in Richmond with the legislature, that was Cofer's terrain. He ensured the 6,000 ship movements a year were safely and efficiently coordinated and that any disruptions were resolved quickly. The port was an economic engine for Virginia.

During my first week, I took a boat ride down the Elizabeth River with Captain Cofer, who described in detail how George Washington, the surveyor, played a critical role in the port's early design. Then he pulled out a copy of a map from a predecessor that was drawn over 100 years ago. It had a diagram of how Craney Island could be expanded into a larger terminal. That expansion had finally begun. Cofer was the corporate knowledge and strategist of the port's growth and potential. It was recently named the fastest growing port in North America.

On the other hand, Bill Burket was a deep-voiced, burly, fire chief who always had five radios strapped to his chest. He was an ideal leader of the blue-collar teams. He was the coordinator of fire, police and EMS.

Both men were smart, but more importantly, they knew how to get things done and worked as a team.

While Sector maintained the port's command center, rarely did I surprise them with a brief. Cofer was already fixing the problems with ship movements, and Burket was usually already on the scene directing a response of emergency services. We had a good team, but November 2009 tested us all.

Just as hurricane season was thankfully winding down, Hurricane Ida formed off the Gulf Coast and headed north. The projected path of the storm showed it coming ashore around Mississippi and breaking up rapidly. In Hampton Roads, the port officials paid little attention to it since we had the Appalachian Mountain range between us. This was a bad move on my part. The storm hooked a right when it made landfall, crossed over Georgia and then reformed rapidly off the outer banks of North Carolina and headed north. Then it collided with a high-pressure ridge from the north. The counterclockwise spinning storm sat stagnate off the mouth of Virginia for effectively three days. We were essentially in a blender with hurricane force winds. The weather guys usually were spot on, but they kind of missed this one.

Normally the Sector would follow a regimented plan that requires port actions at forty-eight, twenty-four and twelve hours out from tropical storm force winds. When it became apparent that hurricane force winds were going to play havoc to the region, we pulled our boats just in time and sent the cutters upriver for refuge. I closed the port. That meant anyone in trouble may not get help. The last responders to pull their boats were the Virginia Marine Police, and they managed to rescue the remaining folks in need. We had not lost anyone yet. Then a series of challenges hit us like a freight train, one car after another.

One of the first calls came from the Surry Nuclear Power Plant up the James River, stating they had several construction barges break free and were now jammed in the cooling water intake. If the Coast Guard could not help them get a tug, the power for the plant would have to be shut down in approximately four hours. I gave my boss at the Fifth District, RADM Wayne Justice, a courtesy heads up, and that we were working it. Remember, don't be the senior man with a secret when the power goes out.

We also got word that there was a sailboat struggling in the weather just up the river from the nuclear plant.

Our team reached out to Fort Eustis, the homeport of my old unit Port Security Unit 305. I knew they had their own tugs and were right across the river from the power plant.

We also contacted commercial tugs upriver in Hopewell, Virginia. Both were granted permission to move in the closed port and responded. The Army provided a search and rescue platform and supported the sailboat, and the commercial tugs yanked the wayward barges out from the power plant. The power remained on.

Just as that wrapped up, a report came in that all five, deep-draft, coal ships at anchor off the port's mouth were in distress. They were in ballast, meaning they were empty, so they had a lot of sail area or freeboard exposed and were now dragging anchor toward the Bay Bridge-Tunnel. That same bridge-tunnel complex that was almost destroyed by one ship on my first day. Now we had five to contend with.

Normally, when ships move within the port, they are required to have a pilot onboard. That would be Captain Cofer's lane. Pilots are experts in driving massive ships from the sea into their port of responsibility. One does not become a pilot overnight; it takes years and years of experience. As part of their qualification process, they have to reproduce all the harbor charts down to depth curvatures from memory.

But in this storm, there was an added problem. Despite being close to the distressed ships, a dredge pipe had broken free in the blender and trapped the pilot's boats from responding. Coached by the pilots over the radio, the crews on all five ships were able to get underway and shifted north to seek a lee from the tip of the eastern shore and better holding bottom off Cape Charles.

The reprieve was short-lived. We received a report that the dead fleet from the Maritime Administration, which was over thirty WWII-age vessels permanently moored in the James River, was breaking apart. This

The dead fleet anchored in the James River (photo courtesy of DVIDS)

permanently moored cluster of ships had taken direct blows from hurricanes in the past and never budged. This time, it was different. The *Monongahela*, an 870-foot-long ship that was in the middle of one of the clusters, had broken free and was drifting south toward the James River Bridge that connects the heavily populated peninsula with the mainland of Virginia to the west. Fortunately, the wind pushed the vessel aground on the river's west bank before directly threatening the bridge's destruction. Mother Nature caused this problem but also helped solve it.

Then came word out of higher headquarters that my beloved Portsmouth Federal Building which was home to both the Fifth District and Atlantic Area staffs had been compromised by flooding. The water was so high that it made getting to the building impossible, and even the emergency generator was reported to be knocked out.

Our higher headquarters went offline until they could stand up an alternate command and control site. I still had cell phone connectivity with key leadership, but they were challenged, as were we. Perhaps not the end of the world, but ironically, they were having an annual flag conference in Virginia that week so all the visiting admirals saw the port carnage play out on TV. Perfect!

Already getting tired, I thought, surprisingly, we had lost no one and minimized property damage.

Then came the mayday at midnight. The 570-foot, unmanned, Crowley barge, reported to be carrying chlorine, had broken free from its tug. It had been on a route from Puerto Rico to Philadelphia and had battled the storm until, finally, the tow line parted. It was off the south-eastern shore and drifting toward Virginia Beach, a city of 600,000 people.

Crowley barge that broke free in the storm

Chlorine is a deadly gas, and the prevailing winds would carry any release toward the population center of Virginia's largest city. The story was picked up by national news. Fortunately for us, it happened in the middle of the night, so we had some time to work on a response. Our first attempt was to sortie tugs to intercept the drifting barge. The port was still closed, but we took a chance. When the tugs got to the mouth of the port, they hit twenty-foot seas.

Having been on the *Key Largo* and trying to get to Haiti when we hit fifteen-footers that almost ripped off the gun mount, it was a no brainer to turn them around. Luckily, the barge on its southwesterly track just missed the huge navigation structure called Chesapeake Light.

The tug that was originally towing the barge had followed, giving us up-to-date position reports. That was about all they could do because there was no way to reconnect the tow. The barge didn't have anchors either, so they couldn't stop the drift.

When we got the original mayday call, we connected with Bill Burket, who had all the connections with Virginia Beach fire and emergency services. I also called our Atlantic Strike Team in Fort Dix, New Jersey. These were the same guys who bailed us out during the satellite mission just a year ago. They jumped in their trucks and headed south with air monitoring equipment.

Don't be the senior man with a secret. I alerted my chain of command and the mayor of Virginia Beach of the unwanted visitor due to arrive at first light. When the barge finally grounded on the beach, we still weren't finished. The bottom was relatively sandy and flat, but Burket and his guys on the beach reported that the waves and wind were hitting it in such a way that it was bouncing its way south, toward the iconic fishing pier at Sandbridge. Like most beaches on the United States east coast, rows and rows of condominiums overlooking the ocean's morning gift stood just above the dune line. It's normally a great view, like my high rise in Fajardo. Unfortunately for us this time, if you went up to the fourth floor, you'd be about eye level with the deck of the HAZMAT barge.

Burket's guys were in the surf trying to rig ladders to board the vessel, but it was dangerous. I called RADM Justice and gave him an update on current events, letting him know we had perhaps a half-mile to the pier. He said he was putting money on the pier getting run over. The winds subsided a bit, just enough to allow the media to venture to the sight. Despite having the strike team setting up air monitoring and passing out shelter-in-place instructions, the media set up cameras that fed national news. They wanted to see the carnage of the barge crushing the pier.

Beyond the pier getting crushed, we were concerned with the footing and damage it might cause to the barge.

We said, not on our watch! So, how do you stop a massive, unmanned barge without anchors moving laterally across the surface of the water? The answer: you sink it!

But the responding team wasn't going to shoot at it like some abandoned drug boat; it had hazmat onboard. Our port team huddled and came up with a plan. I reached out to Captain John Harding, the commanding officer of Air Station Elizabeth City, North Carolina, which was roughly an hour away. Normally we'd go through the district, but they were flooded, and we didn't have time. I told him I needed a hookup—a helicopter—right now to fly to Sandbridge. The wind had just reached the upper parameters of when they could fly, and they were enroute.

Two salvage guys met the helicopter on the beach. With a wide media audience watching live, and about 200 yards from the fishing pier, the helicopter hovered over the barge and lowered the salvage team. They moved quickly and flooded the ballast tanks. About 100 yards out from the pier, the barge stopped!

It was like a movie, and although the media didn't get the footage of the pier getting crushed, they still had some drama. We only had enough time for a few high fives, because we were not done yet with this storm.

We had a cruise ship offshore with irritable seasick passengers eager to conclude a vacation to forget. They were bound for Baltimore. Unfortunately, the Army Corps survey boats, which would generally survey the main shipping channel for debris and storm generated shoaling, had sought refuge behind the intracoastal waterway locks.

The flooding from the storm was so severe that the water level rose high enough to fry the mechanism to open the lock. The survey boats were effectively trapped and unable to check the channel.

By luck, a smaller, 500-foot freighter offshore reported tank damage and contaminated fuel and needed to come in immediately, or they would lose all power. This was a gamble, but Cofer thought if they took it slow, the ship could be safely brought in and serve as the first guinea pig ship to determine if the channel was clear. He was right as usual. Slowly the port was opened, and we even got the cruise ship in shortly after. But like all storms, the weather eventually passed, but the cleanup took time.

As the sector commander, I attended monthly mayor's meetings focused on Hampton Roads transportation along with the Navy's commanding officer from the Norfolk Naval Station which employed 80,000 people. That's twice the size of the entire Coast Guard. We always sat together. She was great. As it happened, the Navy Captain was married to a Coast Guardsman who had been in the port security unit with me.

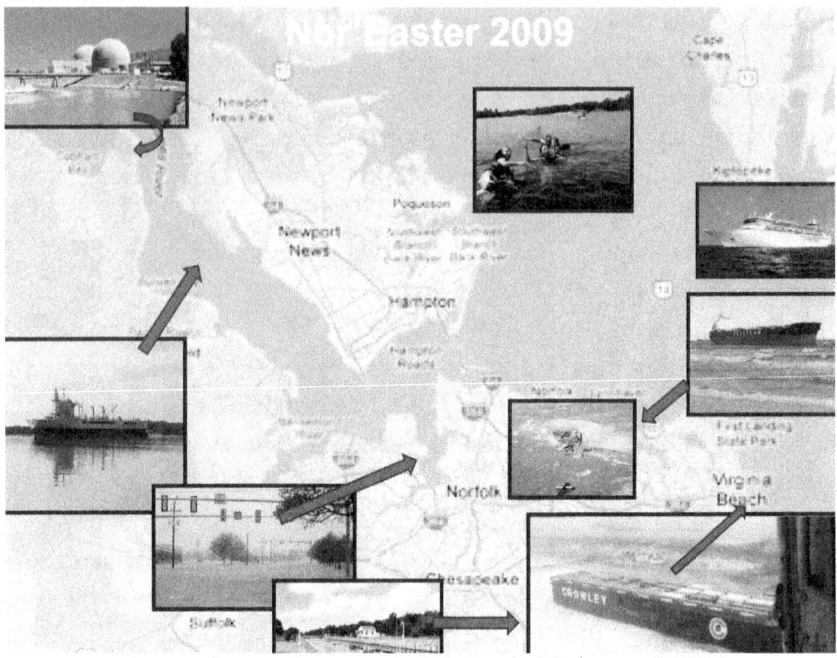

A map of events that transpired during the Nor'easter of 2009

At these meetings, our job was to say nothing, serve as eye candy and eat the snacks provided. With the seven cities in Hampton Roads, they rotated the chair duties, and this month the chair was the mayor from Virginia Beach. Just prior to the meeting, I received the call that the slightly rougher weather had helped break the bottom suction, and three tugs were able to wiggle the HAZMAT barge free, and it was successfully off the beach. I whispered the news into the mayor's ear, and he led off the meeting with that success story.

As it turned out, the barge event made national news and had become a bit of a tourist attraction; it was a boon for the town. We received a very nice letter from the owner, Mr. Crowley, on the handling of the situation. Everything was almost back to normal, but the *Monongahela* took several months to remove. Since then, the Coast Guard and Maritime Administration have continued to send the dead fleet to scrap, and now there are only a handful of ships left.

Lessons Learned:

Hurricanes on the east coast normally track north and have a counterclockwise motion. If you have an opening to the east, expect flooding during the first half of the storm.

Get barges clear from the intake of all nuclear power plants.

Know where the best bottom is for anchoring large ships.

If you have a huge, unmanned HAZMAT barge drifting down on a city of 600,000, notify the locals, then put a salvage team on by helicopter and flood the ballast tanks.

Make sure your post-storm, survey boats are in a safe position to respond.

Large, grounded vessels aren't easily dislodged. You need to wait for the weather to produce some swells to help break the bottom suction caused by mud and sand.

Chapter 34

Train Hanging Off a Bridge, Man in a Raft

I'D BEEN IN THE job long enough to know that when the phone rings after midnight, it's never good. It was a Saturday morning at 0400 when the command center called. The report briefed that a locomotive with a single engineer on board and approximately 100 coal cars was dangling from an open bridge. The bridge was located over the intracoastal

waterway in Virginia and near the North Carolina line. That bridge is normally stored in the upright position until the approaching train communicates to lower the bridge so it can cross.

The rule covering bridge operation over water is that if the maritime

Train suspended off the bridge over the intracoastal waterway
(photo courtesy of DVIDS)

traffic existed before the rail traffic, then maritime takes priority, and the bridge is always stored in the open or upright position allowing maritime traffic priority. The train's engineer never effectively communicated the intention to cross and have the bridge lowered. As the train rounded the turn, the engineer used all the brakes he could and slowed the train but still had enough momentum to send the locomotive off the end.

The locomotive diesel tanks ripped open in the process. The train hung precariously over the water.

I have set default priorities for these calls, because at that time of the morning, it's hard to focus. Those priorities are people, the environment, property and the economy. I just say P.E.P.E. Status of the people is always the first question. The good news was the engineer was uninjured.

This was quite the spectacle, and the footage made it into national news. I used my Captain of the Port authority and immediately closed the waterway so as not to have another vessel potentially hit it, wake our small boats, or drag the spilt diesel into uncontaminated areas. That covered people and the environment, but we also needed to get the mess cleaned up quickly. That meant to repair the train and bridge and get the disrupted economy going again.

Hampton Roads is the nation's biggest exporter of coal, which is mined in the Appalachian Mountains, where I used to hike in Boy Scouts. From there, it's a downhill run for the trains to bring it to Hampton Roads, the deepest port on the east coast. Interesting to note, coal ships draw more water than cruise ships or aircraft carriers, at about fifty feet.

I made it to the scene where Sector Hampton Roads pollution responders and our good friend Bill Burket already had the command bus set up. Of course, the pollution responders took this opportunity to, once again, make a strong case for Coast Guard ATVs for this kind of mission (those guys are always looking for toys).

We made quick work of the cleanup. Thankfully, diesel evaporates rapidly. The railroad brought in a barge, lifted the locomotive and repaired the bridge in a day. Big business can move fast when it means big dollars.

These pop-up crises were easy to handle, and it almost felt like the war on terror and our security paranoia was a distant memory. That abruptly changed when SEAL Team Six, based out of Virginia Beach, got Bin Laden, leading to increased security around the local bases and a request for Coast Guard to provide armed security around SEAL exercises and training. The guys were in our area of responsibility, and as the sector commander, I also held the title of Federal Maritime Security Coordinator.

Terrie Suit, Virginia's Secretary of Homeland Security, was married to a SEAL and let me know that there was a big celebration commemorating the fiftieth anniversary of the SEALs scheduled. With the Bin Laden mission just months earlier, the SEALs had a lot to be proud of! DHS was carefully monitoring intelligence, expecting a retaliatory strike.

On Friday, the sector command center took a report from two unarmed Navy patrol craft who were coming up the intracoastal waterway very near where the train had been hanging and sighted two ATVs on the bank. The two riders, who wore the typical face-shielding helmets, pulled out a rifle and fired in the direction of the Navy gunboats. A few rounds hit the water in front of the lead boat, then a few rounds hit behind the trailing boat. Neither boat was directly hit, but the Navy crews took cover and called it in.

The two shooters jumped on the ATVs and sped off. Our unit was already on high alert, so we closed the waterway down and got the police in Chesapeake and the FBI to respond. The shooters turned out to be two stupid teenagers (not mine), and frankly, they were lucky. If the Navy boats were armed like all Coast Guard boats in the area, they might have returned fire in self-defense.

Several week later, the sun had just come up when a fisherman near the Surry Nuclear Power Plant on the James River spotted an adrift, small, green raft. When he came alongside, he snapped a picture. In the raft was a man who had a chain around his neck, attached with a padlock to a large black pelican case. The man seemed very incoherent but then told

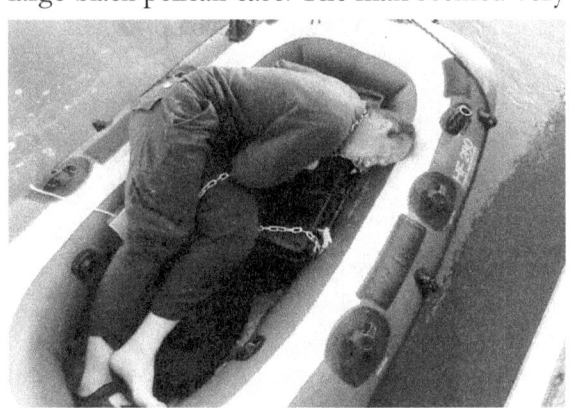

the fishing boat crew to get back and stay away.

That picture went viral. By chance, Secretary Suit was in the area, and I gave her a heads up in case she wanted to come to the command center. About the time she arrived, that picture had made it to national news, so she called the

Man chained around the neck off the Surry Nuclear powerplant adrift in the James River

governor. Meanwhile, we had a couple of big problems to contend with.

The raft was drifting down the river toward the dead fleet. Remember the *Monongahela*? They were covered with sharp barnacles and would no doubt pop the small raft.

There was also other shipping traffic in the area, as 500-foot ships can go all the way up the James River to Richmond, so we closed that section of the river off. We launched Station Portsmouth's small boat with a bolt cutter and told them to pull in briefly to Newport News while in route to pick up a police negotiator. We had no idea what was in the box; it could have been almost anything, including an IED. I'm pretty cautious after my trip to Colombia. We directed the small boat to remain a distance away, unless it was critical to intercept the raft from drifting into the dead fleet.

Meanwhile, we got the state police to send their bomb squad unit to a boat ramp on the west side of the river that was relatively remote. On the scene, our boat crew and police negotiator talked to the man, and they floated him a life jacket. He complied and put it on. Then they floated him the bolt cutters. He was too weak to cut the chain or lock, so boat crew floated him a tow line. The plan was to tow him to the controlled area. The sea swells were out of the east, so it made the most sense to travel the shorter distance west and down swell. Then the FBI returned our call and said they wanted to take the boat east to Fort Eustis.

I told the FBI that ship had literally sailed. We were close to the controlled area. The boat crew passed the tow line to our station truck, and we towed the raft up the boat ramp. The bomb squad came in with heavy bolt cutters and freed the man. Then they used their portable x-ray machine to determine what was in the case.

It turned out to be sand, and this was a failed suicide attempt.

This was not the first time that a search and rescue case became intertwined with law enforcement.

Lessons Learned:

Bridges are normally stored in the upright or open position if the maritime transportation mode existed before the rail.

There may be only one person on a train with 100 coal cars.

Teenagers can be stupid.

Sometimes those contemplating or attempting suicide have a change of heart.

When facing a unique situation, know what's in your tool bag and who can help.

Keeping shipping channels open is critical for the economy. Any disruption has huge consequences. Just in time shipments are the wave of the future. Deeper, wider channels are safer.

Chapter 35

Americas Most Wanted and NASA Scientists Capsize at Night

I STARTED THE SECTOR command job early in April of 2009, which meant the big boating season was just ahead. The Coast Guard always does a big boating safety push in May, but we had a weekend where we had three maritime fatalities before that happened. One was a surfer, one fell off a pier and the third involved two brothers on a small boat outside of the Little Creek inlet.

Shortly after the brothers had gotten their boat underway, one moved to relieve himself off the stern while the other stayed at the wheel. The brother driving reported hearing a splash, and when he spun around, his brother was gone. After searching the area, the man at the wheel called the Coast Guard for assistance. We were all over it, with boats and aircraft responding in minutes. You could not have picked a better spot to get in trouble.

We searched and searched but couldn't find the missing brother. The brother's story sounded fishy, but it was obvious he had a mental handicap. Some things were not adding up. If you're going out on a long fishing trip in an open boat, you usually hit the restroom before you go. They had been underway perhaps ten minutes.

I called my partners in the marine police, and we released a joint, early, public-service announcement on TV from the Sector Command Center. This was in advance of our regular boating safety campaign, stating we had two confirmed fatalities and one man still missing over the past weekend and stressed the need to wear life jackets. That aired on a Sunday, and that news was seen in Virginia and southern Maryland. That Monday, we got a call from Maryland Police stating the missing man had been convicted and was scheduled to be sentenced for a funeral home scam in Maryland.

My gut said this guy was faking his own death. I was not alone, the command center watch, led by LT Ray McKay, had worked plenty of search and rescue cases, and it didn't make sense to them either.

Everything pointed in this direction, and we even had grainy video of someone on the jetties. We reached out to our law enforcement partners, but they didn't come up with anything either over the next few months. Ray and I always thought this guy was sitting on the beach somewhere laughing at us.

The man who faked his death

So, Ray, on his own initiative, contacted America's Most Wanted TV show about six months later. They were interested and helped recreate the whole case for their show. About two days after the airing, there was a tip, and the suspect and his girlfriend were apprehended in Texas. Justice at last. In addition to his funeral home sentence, he had to repay over $100k for the search expenses that followed his disappearance.

Fast forward a year. It was April, yet again at midnight, when a group of NASA PhD grad students were wrapping up a party in Newport News, Virginia. There were six men and four women, all healthy and in their late twenties. They decided unwisely to take a twenty-foot sailboat out on the James River for a moonlight cruise. They were north of the James River Bridge but south of the dead fleet. That point in the river is tidal and approximately five miles across. At that time of the night and time of year, there was usually no river traffic. It was a cool, clear Virginia night.

The ten made it out to the middle of the river when the girls said they needed to use the restroom. You see the theme here? There's no bathroom or even a port-a-potty on such a small-cabin sailboat, so it was mutually decided that the ladies would go to the stern, and the six guys moved to the bow, giving them privacy. That mission was successful but provided foreshadowing on how unstable the boat was for ten people.

Later, one member tried to reset the sail, and the tiller was left unattended. Unlike the previous case, there was actually a real unintended man overboard. The boat capsized, and all ten were plunged into the cold dark river water. No one wore life jackets. To make matters worse, the lifejackets were on the boat but stored well up in the cabin and unreachable in the turtled boat.

They grouped together treading water and planned their next course of action. I want to say this again: these were very intelligent, fit people, despite the factors that got them to this point. They didn't have consensus on a plan. Five said the right thing to do was to stay with the boat and hope to get spotted. These folks were not great swimmers. There were no lights in sight, except for a very distant glow on the horizon to the west. The other five initially saw no chance of rescue because no one knew they went out. They were relatively strong swimmers, so they started the two-mile swim to the west bank using a fender and a floating gas can in hopes of making it to shore and alerting the authorities.

The group who initially stayed with the sailboat realized quickly that if they were not swimming, they would succumb to hypothermia. They eventually abandoned the capsized sailboat and started the swim.

About thirty minutes later, by sheer luck, a tug and barge coming down the river bumped into the turtled sailboat. They called the sector command center on channel sixteen.

"Hey, Coast Guard, just to let you know, we saw an overturned, small sailboat that probably drifted off a bank with the recent heavy rains. We called out but didn't see or hear anyone. Just want to let you know."

There was a young petty officer, OS2 Andrea Cobb, on watch for that call. She could have just noted it as debris, but we have a rule: if in doubt, launch. That's exactly what she did. We had the Cutter *Chock* in the area, and we got the local fireboats from Newport News, as well as a helicopter from Elizabeth City airborne. Andrea also made a radio broadcast.

A few minutes later, the initial five who decided to swim made it exhausted and cold to the western bank of the river. It was a remote area, and the current had pushed them away from their intended target. There is one large home all by itself out on the point, which is normally vacant, but by luck, there was a caretaker there that night. They made the 911 call.

That's when we knew there were others in a desperate situation. A news helicopter overheard the broadcasts and filmed several people from the second group being pulled out of the water. This group, composed of the five poorer swimmers, had briefly stayed with the boat before starting the long swim. It was a very dramatic rescue.

Where the first group swam ashore was near my home, so I was able to intercept the survivors. Even the ones picked up from the water were delivered to the bluff where I was. One swimmer from the second group who had been pulled from the water was undergoing CPR all the way in. We got him to the ambulance and on the compression machine, but they were unable to save him.

A female survivor told me she watched one of the men treading water with her tire and eventually go under and never resurface. "He's dead, he's gone."

We recovered nine of the ten on the surface. Eight of the ten ultimately survived. With divers, we recovered the missing man deceased several days later. I tell this story because these were all brilliant people. Everyone can have a bad day, and the ocean will make you pay.

As it turned out, my neighbor was a sponsor for their academic

program at NASA. Several weeks after this tragic case, she told me a couple of the survivors were having a tough time and wanted to speak to and thank those involved in the rescue. One of these ladies was the one who saw her friend submerge.

Recovered sailboat that capsized in the James River with ten NASA employees aboard

I arranged for them to come to the command center.

They brought flowers for Andrea, who had made the right call to launch and was credited with saving their lives. I also gave the ladies our challenge coins to present to the firemen and Coast Guard crew who pulled them out of the water.

It was a great moment for those rescued and perhaps a greater moment for Andrea and the rescuers. One of those survivors joined the Coast Guard Reserve.

Lessons Learned:

If it looks like a duck, walks like a duck, it's probably a duck.

Sometimes staying with the crash site is not always the right answer. By swimming for help, the first group triggered a rescue and provided vital information about how many people to search for. The swimmers, by exerting energy, stayed warmer in the cold April waters. Every situation is different, and sometimes hypothermia and stress can affect judgement. It's my belief that the Command Center, rescuers and those in the water made the right decisions. While tragic, it could have been much worse.

When in doubt, launch rescue assets.

Even the smartest people can have a bad day.

Grieving can take a long time, but your actions can help others.

Chapter 36

Wargame and Deepwater Horizon Oil Spill

I HAD BEEN TO the Maritime Operational Threat Response war games in the past with my previous jobs. This time I was invited because the scenario had a force attacking the port of Norfolk. This was an international affair. We had Canadians and Mexicans unified with the United States defending North America. As a captain, I was one of the more junior members present. This year, the war games were being held at the Naval War College in Rhode Island.

Just as we were getting into the war game, cell phones started ringing. The Deepwater Horizon oil rig had exploded in the Gulf of Mexico. There was a loss of life, and the spill-preventing device, almost a mile down at the sea-bed, had malfunctioned. This was a catastrophic event.

At least the Exxon Valdez spill in Alaska was limited to the quantity on the ship. The Deepwater spill seemed to have an unlimited amount of oil pumping into the gulf with

Fighting the Deepwater Horizon oil spill
(photo courtesy of DVIDS)

no way to stop it. The admirals departed for a real-world scenario and left us back-seaters to wrap up the war game.

President Barak Obama and Admiral Thad Allen (photo courtesy of DVIDS)

One of my five hats as a sector commander was to serve as the federal on-scene coordinator for oil and HAZMAT response in the coastal zone. As part of that role, I chaired an interagency/interstate committee with many partners and published a comprehensive plan on dealing with all sorts of threats. It's detailed to the point where there is a grid system that identifies what species were present and oil containment and clean-up strategies. We also conducted frequent exercises to refine and improve our plan.

I didn't expect the oil spill south of New Orleans to directly impact my area of responsibility, which was well up the east coast. Like I had underestimated Hurricane Ida in a similar location, I was partially wrong.

The Deepwater Horizon oil spill almost broke the Coast Guard. Our sector sent three-quarters of our spill response supplies in Virginia and Maryland south to fight the spill at its source. In addition to supplies, we also sent about 200 people from the command. The problem was unprecedented.

Many in the public, academia and media couldn't understand why the US military, with all of its might and resources, couldn't solve this problem independently. There was a lot of fear that this could never be capped, and general ideas to solve the problem were flooding in. Some were less than helpful.

Shortly after returning to Virginia from the war game and resuming command, I received a fifteen-minute warning before having to do a live TV interview on Virginia's oil spill readiness. I was starting to get in my groove with the media, so that went fine. Then I got the call from Governor O'Malley's office in Maryland stating the governor wanted to meet me in Ocean City to discuss our response efforts.

Ocean City has a billion dollar-a-year tourism industry. Much like I had seen in the Virgin Islands, the reputation for a safe and enjoyable beach experience is critical to the area's economic health. I had technically been in command over that area for six years since it included my time as a group commander, so I was relatively comfortable with the area and the local responders. If the beaches were shut down, no doubt it would be devastating to the state. Plus, if the oil made it into the Chesapeake Bay on the backside of the Eastern Shore, it could harm the blue crab industry, the state's symbol.

Having learned my lesson with the Delaware senator years earlier, I reached out to some of our folks fighting the spill on the Gulf Coast and my right-hand man, NOAA Scientific Support Coordinator Frank Csulak. When you think of a scientist, you probably envision a person in a long white coat with thick glasses and a bunch of beakers. That was not Frank Csulak. Frank could easily fit into a biker gang if he wanted to. In fact, I think he rides a Harley.

He was a very smart, dedicated, public servant who knew his stuff. He provided just-in-time data to give me confidence going into the meeting. I also reached out to my Maryland and Virginia state counterparts in pollution response. For Virginia, it was very easy. John Settle from the Samurai jeep ride in Turkey was now the sector's senior reserve officer and the state's lead pollution response guy. I had the honor of promoting him to commander on the Cutter *Eagle*. Alan Williams was the state coordinator in Maryland. He was a great guy, but I had only spent time with him during meetings and exercises.

Based on conversations with Frank, John and Alan, I felt reasonably armed for a discussion with the governor.

When I arrived at the restaurant, it was like deja vu. The governor sat at the head of a long, outdoor table. My seat next to him was open. My counterpart was there as well. Captain Mark O'Malley was responsible for the sector north of my area, including the DC area and the northern Chesapeake Bay. I covered the ocean coast of Maryland and the mouth to the Bay, so this was really my threat to deal with. Despite sharing the same last name, I don't think Mark and the Governor were related.

Along with the governor and Mark, it looked like the governor had his whole cabinet seated around the table. Circling the table was a media ring with TV cameras focused on the discussion. I knew this was a great opportunity to screw up my career once again.

But I had a plan and was prepared. I told the assembled Maryland team we had hung a sensor on the sea buoy at the Chesapeake Bay's mouth, and we had not seen any pollution so far. We checked it regularly.

We had drafted a ship decontamination plan based on a successful tactic used on the gulf coast, in the unlikely event the oil reached this far north. By having that plan in place, we would not necessarily have to shut ports down. I also mentioned we sent hundreds of our people to the gulf, getting hands-on experience with oil spill cleanup. I had a good message. My counterpart in North Carolina had a good quote as well, which I stole. "You are more likely to find dog poop on a North Carolina beach than oil from the Deepwater Horizon Spill."

That's basically what they wanted to hear. Their beaches and tourism remained open through the summer. Governor O'Malley later inspected oil containment booming operations and asked for a full accounting of all the spill response equipment remaining in his state.

A month or so later, my wife and I attended a dinner function with Governor McDonald of Virginia, and we briefly exchanged words on Virginia's readiness.[1]

This experience was educational. If you want to know how someone will react in a crisis, you should first put yourself in their shoes and see what's motivating them. The media typically wants a sensational story because it sells airtime, which translates to advertising revenue. They have a saying, "if it bleeds, it leads."

Politicians—many of which are fantastic public servants—are in the business of getting reelected. Shifting responsibility and blame is step one. Step two is showing you're working hard to hold those responsible for the crisis accountable.

Fortunately for me, my only bad experience was with the Delaware state senator early on. That was not necessarily the case for my counterparts responding to the spill in the gulf.

Two of my friends deployed and served as tactical commanders for the Deepwater Horizon Oil Spill. Both had extensive spill-response experience and were probably the best the Coast Guard had to offer. Roger Laferrier had been the deputy sector commander in Hawaii during the Superferry case, and Merrie Austin who was on the DOG staff with me. She was currently the sector commander in Philadelphia but was hand-picked for this mission. These two rock stars went down and worked incredible hours, and both ended up with medical, stress-related,

[1] Governor O'Malley ran for president but lost in the primary to Hillary Clinton.

heart issues. I'm told that Roger, who retired, had multiple lawsuits issued against him over this particular response. Merrie made flag and commanded the Fifth District.

These huge events are so complex, and someone must lead them. Decisions must be made, many times without all the information needed. It's frustrating when lawyers come out years later, after studying all the facts in hindsight, and bring lawsuits against those who actually helped prevent a calamity. That demotivates great people from taking these jobs with great responsibility. I know it's a way to make a living for those guys, but I hope it's hard for them to sleep at night. Hopefully I won't get a lawsuit for saying that.

Upon their return, our deployed folks recapped an interview with a certain, well-known, national-media anchor who showed a picture of an oiled pelican while on camera and said we weren't doing anything about it. When the camera was off, our folks asked for the location of the bird so they could respond. He refused to give it to us.

Another media outlet offered $500 for pictures of Coast Guardsman relaxing at a hotel. Our personnel worked twelve hours on and twelve hours off, and unlike the private sector, they didn't get overtime. I will not forget that, nor the efforts of so many to protect our coastline.

Lessons Learned:

If agencies screw up big enough, they change their name and get more funding.

Despite the Coast Guard moving heaven and earth, many gulf coast politicians and media outlets wanted to paint them as the villain.

Admiral Allan, named by President Obama to lead the national response to the oil spill, can effectively communicate complex issues with both Harvard PhDs and grammar school-educated fishermen.

The likelihood of oil reaching the mid-Atlantic was extremely small, but governors well up the east coast were very much involved.

The use of dispersants was highly effective yet also very controversial. Academics could not agree. Lock them in a room and tell them to give you one answer.

Protecting the environment rivals all our missions.

Chapter 37

War of 1812, USS Cole and a Surprise

AS MY THREE YEARS in sector command were coming to a close, we had a bit of a finale in Virginia.

After 9/11, we were very security conscious and enforced 500-yard security zones around naval vessels. The year before the 9/11 attacks, the USS *Cole* was at anchor in Aden, Yemen. Terrorists in a suicide boat attack struck the ship near the mess deck, killing seventeen navy sailors and injuring another thirty-nine.

The crew valiantly saved many of those injured and the ship. Now repaired, the USS *Cole* is homeported in Norfolk. On the tenth anniversary of 9/11, the city and the Navy wanted to conduct a remembrance

USS Cole, survived the attack in Yemen in 2000 (photo courtesy of DVIDS)

ceremony using the USS *Cole* in the Elizabeth River as a backdrop. Many senior VIPs and thousands of spectators were expected on the riverbanks and in small boats. One of the hats I wore as sector commander was the federal maritime security coordinator. I had nightmares of a second attack on the *Cole* during the ceremony. What a spectacle that would be.

After all, terrorists go back and try to finish jobs. Just look at the 1993 bombing of the World Trade Center. That attack proved fatal but didn't bring down the towers. Fast forward to 9/11, they changed tactics, brought the buildings down and killed almost 3,000 people.

But the tenth anniversary remembrance went as planned. It was good to show our adversaries that we can take a punch, and we're not scared. This was a good ramp up for the 200th anniversary of the war of 1812.

Most people remember little about this war. It was round two with England. The United States was still a fledgling country, and England was still a military superpower. The English fleet sailed into Virginia and up the Chesapeake Bay, landed an army, and even burned our capitol in Washington, DC. Assholes!

The naval forces played a significant role in the war, including action by the Revenue Cutter Service created by Alexander Hamilton in 1790 and was the predecessor to the US Coast Guard.

Harborfest is one of the biggest maritime events on the Atlantic Coast and incorporates tall ship parades, concerts and fireworks. These are held annually, but Harborfest 2012 was even bigger as it incorporated the War of 1812 celebration. The lead ship for the main tall ship parade would be the Coast Guard's tall ship *Eagle*.

One of my best friends and cadet travel partner was J.J. Kauza. He mentioned he was coming to Norfolk for Harborfest with his two daughters to show them what he had done some twenty-five years earlier. His time with his teenage daughters was precious. J.J. had left the Academy before graduation to be a rescue helicopter pilot on the west coast.

Because I was the captain of the port and had a friend who was the commanding officer of the *Eagle* (now-Rear Admiral Eric Jones), I was able to not only get JJ and the girls a tour; we got them on a boat ride out to *Eagle* and great seats on the lead ship in the parade coming into port. There were perks to go along with the challenges of the job.

As the sector commander, not all perks are exciting. I had many dinners and functions at night and on the weekends. Nye and the kids rarely wanted to attend these events, and I certainly didn't blame them. My negotiated request was to have them along for just a couple events a year.

Every year, we had an annual Virginia Maritime Association dinner. It is a big deal with probably 1,000 folks representing the 550 maritime businesses. It was a time to get the group together with government partners to celebrate the port and all it does for Virginia. It was a formal affair with a jacket and bow tie. They had me at the head table of dignitaries that included captains of industry, Navy, Army and political leaders.

It was my third year, so I knew several of the folks at the table. It was similar to the mayor meetings where you eat, nod your head, clap at appropriate times, and most importantly, don't say anything. As I was getting settled, to my surprise, a little girl approached me. It was my daughter Storm in a nice dress! I was shocked.

"What are doing here?" I had literally just left the house with the whole family in shorts and t-shirts.

She waved and went to sit in the audience with my wife, son and that dirty, rotten, scoundrel Captain John Little, the Sector Deputy.

I received a very nice award called the Port Champion on behalf of the Coast Guard presented by none other than Bill Cofer. John, as my deputy, had completely managed the whole thing, including sneaking in after I departed my house and driving my family to the event.

As for John Little, I could not imagine a better deputy. During our tour together that included the Haiti earthquake, he had deployed on short notice and served as a captain of the port in Port au Prince, getting in relief supplies and working to get the waterway back open. That included using the Hawaii Superferry which had been in mothballs in the Port of Virginia after its failed mission in the Hawaiian Islands. I would ultimately sign its new certificate of inspection to allow its critical lifesaving mission to that devastated island of Haiti.

Later John deployed as the executive assistant to Admiral Zukunft (from St. Croix fame) who directed the later stages of the response to the Deepwater Horizon oil spill response. John fleeted up to relieve me in command, but I wasn't done with that rascal.

Lessons Learned:

Make sure you not only set up a small boat perimeter around the
ship but also sweep the pier face for diver-planted explosives.

Have snipers on the building roofs, just in case.

When you can't possibly cover every threat vector, let the public
help you identify threats. Remember the story of the sheep,
sheep dog and wolf. Ninety percent of the people in the world
are good, and they can be additional eyes and ears to alert law
enforcement of suspicious activity. You need to ask them for
help.

The British, two hundred years earlier, sailed up the Chesapeake
Bay, landed an army and burned our capitol during the war of
1812.

Reputations change: The Superferry, originally a negative symbol in
Kauai, improved their reputation significantly as she sailed with
earthquake relief supplies to Port-au-Prince. No protesters this
time.

I would have been screwed without John Little as a deputy.

Chapter 38

Boating Accident and a Great Deputy

JOHN LITTLE HAD STARTED as a seaman and made it to captain.[1] I couldn't have asked for a better deputy. I was more a surfer type who loved to go on the scene during a crisis; he was more of the respected father figure. (He was indeed slightly older than I was.)

When the Haiti earthquake occurred, thousands of lives were at

VADM Rob Parker, Captain John Little, Congressman Scott Rigell, CAPT Mark Ogle, and RADM Dean Lee touring the port of Virginia

stake. Port au Prince was severely damaged, limiting its ability to take in humanitarian supplies. It certainly brought back memories of the policy of returning fleeing migrants. These poor people can't buy a break.

[1] John Little was once roommates with another one of our partners in crime, RADM Keith Smith.

Atlantic Area was searching for someone to go down and serve as a captain of the port along with the Haitians to make this work. John Little was their number one choice. Of course, John asked me, in typical John fashion, if it would be okay to support the mission. Despite being my number two in command at the Sector, this was a no brainer; thousands of lives counted on him, and he had trained a great staff to fill his position as deputy sector commander. The answer was "go and we'll figure it out." He was in the air in hours.

It wasn't that the Haitian authorities couldn't do it, but they had been in the epicenter of the quake. The Haitian man in charge of the port was living in his yard and had lost loved ones. John's competence and perhaps compassion for those in a crisis made him the absolute right choice for the mission. He did an incredible job, and no doubt saved lives by his actions. Six months later, he was requested to serve as an executive assistant for Paul Zukunft (a future commandant), who was taking over as national incident commander role from Admiral Allen for the Deepwater Oil spill in the gulf. John once again deployed and excelled.

Not forgetting that he had surprised me with the award ceremony, it was payback time. The deputy at any sector is plugged into everything. It's extremely hard for any detail to get beyond him, especially John Little. We were in our traditional morning operations brief in a separate room next to the sector command center floor. The curtains were always drawn to limit distractions. The command center was a fairly large room with normally five to ten people on watch. Its walls were adorned with TV monitors, camera feeds and charts. I had set up a code with the watch stander to make up a reason to interrupt our brief.

This was shortly after the time when the tsunami hit the Japanese coast, and the nuclear power plant was compromised, so that was the story to pull us out of the smaller briefing room and onto the command center watch floor. When the door swung open, there was John's lovely wife Michelle. (I never understood why someone that beautiful would have married John Little, but she did.) Next to her were VADM Rob Parker, the Atlantic Area commander, and RADM Dean Lee, the Fifth District Commander. Behind them, stuffed in like sardines, were a hundred or so folks from the sector who had all crammed in silently. It was a successful exercise in operational security, and it was a total surprise to John. VADM Parker said some nice words and presented a well-deserved award to John for his efforts in Haiti.

By mid-summer, it was time to turn over the reins of command to John in a big outdoor ceremony. We probably had a couple of hundred guests, including Congressman Bobby Scott, Virginia's Secretaries of Department of Transportation and DHS and a host of agency heads. It was a bit of a who's who in the port of Virginia. My old boss from Yorktown, RADM Steve Ratti, now the current Fifth District commander who had relieved Dean Lee, presided

RADM Steve Ratti presided over CAPT Mark Ogle's Change of Command

over the change of command. I must admit, I never liked ceremonies, especially those I had to speak at. But I was ready.

With the VIPs, great partners, family, and friends looking on, I said they had all been very instrumental in the port's success over the past three years, but I wanted to recognize two individuals in the audience specifically. You might be thinking it was Bill Cofer and Bill Burket. That would have been expected. Perhaps the head of ICE Mike Lamonea, CBP Mark Laria, or TSA's Jeff Horowitz. No, even though they deserved it. There was some brief looking around, and then I called up a thirteen-year-old boy and his nine-year-old sister. They were sitting toward the back near my classmate Gary Thomas who gave me a thumbs up. And no, they were not my kids.

As they walked to the stage, I explained that these two kids had been on a speed boat on the Rappahannock River. It was dark, around 2200, and they were in the bow of the boat with their halfsister and a young teacher who was getting ready to get married. In total, there were ten people in the boat; the rest sitting toward the back of the boat. They were traveling between two campsites on the river and were moving at thirty-five mph when they hit a fixed aid to navigation structure in the middle of the river. The ice-reinforced structure never moved on impact. The collision launched the boy Chase, his sister and his half-sister into the dark water. The schoolteacher was in the middle, struck the structure and was killed instantly. All the adults in the boat suffered severe and incapacitating injuries. Fortunately, the man driving, despite having his face partially filleted off by the windshield, got the call out for help. Coast Guard, auxiliary, Marine Police, fire, EMS, sheriff's office—many of whom were in the audience that day—were all racing to help. They were in a remote area. The Thomases in the auxiliary were first on the scene. It was carnage.

Sometimes, we need help from those in trouble. Young Chase had a broken femur on one side and a broken wrist on the other and struggled to stay afloat in the pitch-black water. But then he heard his sisters crying

for help. They also had broken bones and were panicking. He knew he had to save his younger, injured sisters, despite his own injuries. If he did not, who would? He got both sisters and himself back to the boat and on board.

Milford Haven Boat Crew received medals for heroic rescue

That night, we saved nine people out of ten people, and I only had to do one next of kin notification, thanks to Chase! After the citation was read, we pinned a medal on him, and then I nodded to his sister, who went up on her toes and kissed him on his cheek. Then I pulled out a $20 bill and gave it to her. I mentioned to the audience that I had started the negotiation at $5.

After they sat down following emotional applause, I explained that when anyone in the sector departs, I do a debrief and ask them for the name of

FOREVER CONNECTED!
Chase

CAPT Mark Ogle and
Chase Buchanan

someone who flies under the radar, who was especially helpful and a great shipmate. They can only pick one person. When I got that name, I send that person a note and copy their entire chain of command. I also ask if they had a former commanding officer who did something particularly great that I might copy in the future—like the thing I just mentioned with giving the kid an award. I got that from Dean Lee. It is always good to get fresh ideas if you're an aspiring leader.

I also ask if there was something I do that was particularly bad. Usually, when I ask that question, the member is looking around for hidden cameras. Once, when I asked that of a retiring command center chief, she didn't say anything during our session, but she returned twenty minutes later with a large coffee can carefully wrapped. On the wrapper, it said, "Captain, please, there are sailors present."

That was the start of my swear jar, which I filled up, perhaps a couple of times. My parting gift was to present the jar to our great Command Master Chief Jim Labarre to place toward a crew's pizza party (there was plenty of money).

It had been a good tour, but I would be less than honest if I told you I wasn't happy to turn in my command cell phone. At any time of day, including weekends and holidays, a ringing phone could mean I'd have to do a next of kin notification. It was the burden of command and that now shifted to my good friend John Little.

It was also nice that the assignment officer allowed us to stay in Virginia. No move was necessary. For my next assignment, I only had to drive about ten minutes more to work at the Atlantic Area.

Lessons Learned:

When possible, always recognize young people for acts of valor. Parents had done that for me with the Boy Scout yellow jacket rescue. This was my time to pay it forward. Chase became a military officer.

Make sure if you are doing something emotional early in a speech, have something funny right after it.

If you manage a large organization, make it a point to get out to the far reaches to support the whole team.

Look for opportunities to mix your team by doing something both fun and constructive. We took our wardroom (junior officers) for a camping trip each summer to one of the three lakes where our auxiliary volunteers manned search and rescue detachments. In the morning, we landscaped, built-out piers and painted classrooms. In the afternoon, we jet skied and ate a pizza dinner with the volunteers.

When in doubt, go on scene. If an accident happens on a demarcation line of jurisdictions, sometimes exerting authority is the right move. Fill the leadership vacuum, especially if lives are in danger.

Our senior enlisted often have more command experience than our officers. Master Chief Heath Jones, who commanded one of the sector's patrol boats, also coached our sons' high school golf team, and became the Master Chief Petty Officer of the Coast Guard. Our service's most senior enlisted leader.

Chapter 39

You Can't Retire

PERHAPS ONE OF THE hardest decisions anyone has to make in life is whether or not to retire. In the military, that means after twenty years of service, formally submitting a retirement letter. At that point, after a particularly bad day, many people have drafted up that letter and have it sitting on their desktop to remind them they have options.

The military does a nice job of teasing you to stay in, though. Promotions mean more money, and after twenty years, there is a 2.5 percent increase in retired pay for every additional year.[1] It's tough to just walk away. The military invested a lot to get people to this point, so it's a tradeoff. For full-bird captains or colonels, the option is to retire or stay in and compete to be retained as a senior O6 and possibly compete for admiral or general. The latter is an extreme long shot. All sorts of stars must align from performance to education to the needs of the service.

But many officers in the service run this marathon and feel they are at mile marker twenty-five and must complete the mission. I'll say this delicately: those are generally officers you'd prefer not to work for. The lion's share will not wear stars, and many depart the service unfulfilled. The corporate world buys out senior executives when they become too top-heavy or want to go in another direction. In the military, captains have locked in a great retirement, so there's no severance package. But the organization must still thin the herd at the top, so those below have the space to promote.

At the twenty-six-year mark, 50 percent of the captains remaining

[1] That has changed under a new system.

292

are screened and not continued for service. That means they will continue to work for a year and then must retire at the twenty-seven-year mark. Those in the 50 percent selected for continuation are allowed to stay on for four more years until the thirty-year point and compete for flag officer. At thirty years, if not yet picked up for admiral, they must retire. I had the great fortune to serve with many flag officers and post-continued captains. They are considered the epitome of patriotic government servants. Many could get out and make easily three times as much in the commercial world but choose to continue to serve.

So, at the end of a great tour commanding the sector and a very dynamic career, I found that I had checked most of the blocks to compete for continuation. I even had a few flag officers who encouraged me to remain in and made calls about service as a chief of staff. I thought the title of admiral would be awesome. It does have a nice ring to it. I knew the likelihood of being selected was incredibly low for me. In the unlikely chance it did happen, the job would probably be horrendous. I had flashbacks to my executive assistant job.

When Jeff Hathaway was a two-star, he explained this to me. If you get picked for admiral, you essentially sign up for five years. Your second star is almost a guarantee, like going from the rank of ENS to LTJG. It's a great honor to be selected for flag, so part of the deal is you must be ready to go anywhere and do any job.

By this point, you have established yourself in certain specialty areas, so you'll have a good idea of where you'd land.

Then I considered my wife Nye, who had gone from being a military brat to a military spouse. She was tired of moving. She's not one of those wives who likes to wear her husband's shoulder boards either. She didn't enjoy the social events which, if I stayed in, would not only continue but increase. Frankly, I wasn't much of a fan either of these events.

The kids were in high school. A move for them would be dramatic. My mom was a widower, and Nye's folks were only thirty minutes away. Thus far, move decisions had really everything to do with me and my career; my family was dragged around. Therefore, I made the difficult decision while at sector to submit my retirement letter.

As a single income family, I had lined up my ducks for a civilian Coast Guard position. I had made it through the process, including interviews, and I thought everything was tracking. Unfortunately, the civilian hiring practice was very slow, and at the same time, I was being shopped for other captain jobs. Then I got a three-way call from the chief of staff from the Fifth District and chief of staff from Atlantic Area. My request to retire was denied. I was to report to the Atlantic Area for the future operations job.

Wholly crap, I had managed to piss off my old and new boss.

I was somewhat devastated, as I had mentally gone through torture to decide to submit the letter and compete for the civilian position. But it was really no one's fault; it was a timing thing. On the positive side, the current chief of operations at Atlantic Area was Captain Steve Truhlar, who had been my classmate. I also didn't have to move, and we had a great team. But it was back to the dreaded Portsmouth Federal Building, where I had already served a four-year sentence. I, of course, saluted and chalked it up to timing and bad luck. The civilian position where I intended to go was filled by the runner up who honestly was a better fit for the job.

Going back to a big staff had its challenges. When Steve was gone, as the senior captain in operations, I sat in hot seat. It was busy! But about a week into it, I was called by the Atlantic Area chief of staff that VADM Parker needed to see me. I thought nothing of it, but when I went up, he sat me down and told me I wasn't continued. What that means in military-speak is "you're done in a year."

I had to give that kind of news periodically to those in my command. It was incredibly hard to deliver; the only thing worse was next of kin notifications. Now, for the first time, I was in the receive mode. I had put in my letter to retire, and it was denied. Now I felt like I was effectively being fired. I apologized to VADM Parker, because he had no role in it other than having to deliver the news. He was a very classy guy and later conducted my retirement ceremony. That said, I took the rest of the day off to get composed. They always delivered this news on a Friday, so we had a saying: if you 're in the selection zone, don 't answer the phone. No news is good news. No phone call generally means you made the cut.

The one rule of thumb I'd always used throughout my career, from cadet days on, was to stay positive and keep earning my paycheck. I knew that when it came to retaining only 50 percent of the remaining captains, all were superstars. The fact that I had signaled I was ready to retire made complete sense to the board.

There were about ten of us in the same boat at the Atlantic Area. Someone brought in a movie poster from *The Expendables*. It had all the

CAPT Mark Ogle as Sylvester Stallone in unit rendition of
The Expendables

action stars lined up. One of the guys took pictures of our faces and taped them over the characters. I was Sylvester Stallone. We put an X over the guy when he departed. We were a team, and when we came across jobs that we didn't want, we shared them with the broader group. The news of non-selection or passed over feels like a kick in the junk, but it actually becomes liberating and exciting after the initial shock.

That said, I knew of a position being created at Yorktown teaching the sector command course with Larry Brooks. I had a killer endorsement from RADM Steve Ratti, but I still worked another year before my face got its X on the poster.

Lessons Learned:

The speed of technology and the 24-hour news cycle means admirals and captains can't allow staffs to develop courses of action as they had in the past. This cuts out valuable training opportunities for junior personnel and is crushing our senior leaders.

Career decisions must include the entire family.

Always be ready for bad news; never think you are a shoo-in for any job or promotion. Sometimes bad news is really good news, you just don't know it yet.

If you are leaving the military, start a year out. Make at least five copies of your medical record. The Veterans Administration will be reviewing it, and it takes up to a year to get a rating.

Chapter 40

Superstorm Sandy and HMS Bounty

I HAD A YEAR before I was officially done. One of my jobs at the Atlantic Area was managing major incidents, especially when it involves more than one district. Upon reporting in I went right to work as Hurricane Issac had pummeled the gulf coast. I needed to track down an old friend for help, Brad Kieserman. I tracked down Brad who had retired from the Coast Guard and was aggressively sought by numerous agencies. He was currently filling the role as FEMA's chief counsel and riding with the administrator during Hurricane Isaac recovery operations. Our Gulf Coast sector commanders had barges and cylinders of chemicals washing down the Mississippi and beaching on both sides of the Mississippi river. Obviously, these were hazards to be addressed.

President Obama declared it a disaster under the Stafford Act meaning the federal government paid 75 percent of the clean-up cost, and the states paid 25 percent. The problem was Louisiana pointed to Mississippi and vice versus on who should pay the 25 percent. That's when things stalled in the cleanup, which is a Coast Guard-led responsibility. Brad, as usual, worked with the states and unclogged the jam.

Hurricane Sandy, or Superstorm Sandy as it was called, raked the Seventh, Fifth and First Districts as it moved north. The entire federal government was spooled up with FEMA leading the whole-of-government response. This was not the FEMA we saw during Hurricane Katrina. They were a well-oiled machine. All eyes were on the news, which painted a possible disaster for New York City.

During one FEMA video teleconference, President Obama made a brief appearance to motivate the federal team. It was an election year, by the way. The Coast Guard is lead on Maritime Search and Rescue as well as pollution response but played other roles, like opening the seaports to recovery efforts. The HMS Bounty was a replica ship used in the movie by the same name starring Mel Gibson. The crew was heading south, mistakenly thinking they could skirt the storm to the west. They were wrong. They got off a mayday that they were sinking right in the path of the storm about 100 miles south of Cape Hatteras. Sector, air station and district responded with helicopters. At Atlantic Area, I had launch authority for a cover C-130. The C-130 provided cover in case the helicopters doing the hoist had a problem. They were outfitted with long-range communications and could launch a life raft if needed. The weather was so rough that a windshield in the plane cracked. That's why we pay these aviation guys the big bucks. This effort saved all but two crew members.

I briefed that as part of our situation report to Secretary Napolitano. President Obama in the State of the Union address that year quoted the Coast Guard survival swimmer who swam up to the Bounty's life raft and asked, "Does anyone here need a lift."

The Coast Guard is very agile, and we can move assets without a FEMA mission assignment. We saw that with Hurricane Katrina. The Department of Defense is very capable, but policies and laws slow their movement. Domestically, we're usually first in and have to hold it together until big brother can roll in in a couple of days. We eventually coordinated with a Navy three-ship surface action group to support post-storm efforts in New York City.

As part of my duties, I also video teleconferenced with USNORTHCOM whose commander was an Army four-star general who I had met when I flew by C-130 to Colorado for a separate mission. During the workweek, normally one of the two admirals—the Atlantic Area Commander or the Deputy Commander—were on the video teleconferences, but occasionally they put in the fourth string on the weekend.

When briefing a four-star, it's best to be very brief and to the point. These sessions are very choreographed with a facilitator normally on the senior member's staff. When it was my turn, I briefed on the Coast Guard's status and asset movements. The general recognized me from a terror exercise and found it amusing I was working a storm too. I said we were a small service.

On that particular call during the recovery phase in the aftermath of Hurricane Sandy, it was mentioned that New York City was in dire need of pumps to dewater the flooded Brooklyn Battery Tunnel. The commander asked all participants on the call if we had any in the inventory to support such a request. My initial thought was of the small P250 dewatering pump we had on the cutters, but then I remembered that all three strike

teams had massive pumps designed for their oil spill response missions. Sure enough, we had them; one was in Fort Dix and close to New York City. Another one was in Mobile, Alabama, which was loaded on a flatbed and was in route. The third

Superstorm Sandy flooded NYC's tunnels.

was in Novato, California. I was able to pass on that good news. For the California pump, I mentioned we could use some help getting that moved quickly. They said to get it to the airbase, and they would get it to New York.

CDR Eric Doucette, commanding officer of the Atlantic Strike Team, was already on scene and working on a plan with the Army Corps to put these pumps in series. It was critical to dewater these tunnels as they were a lifeline for ambulances and other emergency services. Our plan worked. It was a testament to how well disparate government organizations can work together to accomplish daunting tasks in emergencies.

After working six straight weekends, we finally reached the end of an active storm season. There was yet another opportunity for travel, and I was just the guy for the job. Croatia had stood up a Coast Guard five years earlier and were interested in having an assessment done by the world's oldest Coast Guard. The Croatians were also having a growing refugee problem coming from Syria and Northern Africa. It was certainly like what we had seen from both Haiti and Cuba. There were also drug organizations capitalizing on the previous instability in the region and attempting to set up trafficking routes to feed Europe's demand, not unlike Central America.

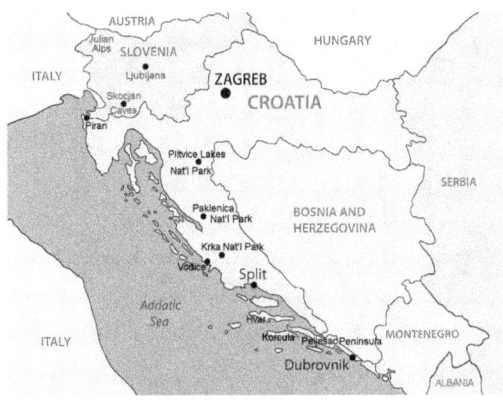

I met up with a travel mate from the Maritime Law Enforcement School, and we spent a week with the Croatians. We not only did tours and presentations but also got underway on their patrol boats to see how they operated. The country was absolutely beautiful. Clear blue water reminded me of the Greek Isles. Lying north of Greece and just across the Adriatic Sea from Italy, Croatia was home to numerous magnificent ruins.

They had also been embroiled in the recent Balkan War. The United States used its influence to stop Serbia's open hostilities. The Croatians we spoke to said they had been close to achieving the desired end state, and the United States stopped them in their tracks. The Balkans were complicated. I had done the Danube River smuggling mission into Serbia two decades earlier, so I had a sense of the long history of conflict and distrust of neighbors.

I have visited a lot of countries, seventy-six and counting to be precise, and it's quite eye-opening when you hear from their military what role the United States played or should play. In most cases, they enjoy the financial support but would prefer it came with fewer strings attached. Of course, that's diplomacy. They knew these decisions were made way above our pay grades, and we were just the ones who carry out the missions.

Upon returning, my staff got involved in trying to resolve a seventy-year-old mystery. During World War II, a Coast Guard aircraft was believed to go down in Greenland's snow fields. With recent global warming, the ice cap had receded, and there was a possible satellite sighting of a missing World War II Coast Guard aircraft. This wreckage investigation required a C-130 with specialized equipment. Flying in extreme latitudes is tricky business, because if the plane has a mechanical or weather problem, the crew is camping among the polar bears. As the future operations officer for Atlantic Area, this mission fell under my control to manage.

The investigation found evidence which was turned over to a special Army team that goes in to recover remains. That was an ongoing mission as I transition to retirement.

In addition to uncovering secrets in the melting ice, the warming planet also affected endangered species that migrated too far to the north. We used Coast Guard C-130 training flights to fly manatees and endangered turtles to the warmer waters of Florida. The rise of the sea level and the

Stranded manatee

increase ferocity of storms also led to more and more migration issues. We saw that in the Caribbean and the Philippines as well as Croatia.

One of my last major operations before retirement was being on a critical incident call, which included the entire chain of command in the Coast Guard and the DHS's command center. There had been a bombing at the Boston Marathon. My role was somewhat limited to transferring deployable special operations teams to ride the ferries to prevent the terrorists from escaping while protecting passengers.

While the Coast Guard didn't play a big role in it, one of the terrorists was found hiding in a boat. Of course, it was in a backyard.

Lessons Learned:

Don't sail into a storm if you have a choice.

The Coast Guard has big ass pumps capable of dewatering flooded tunnels.

Terrorists want media attention. They like multiple prong attacks. Boston Marathon had two bombs. The initial explosion was at the finish line. When first responders and media came in, the second went off. Have your head on a swivel.

Regardless of the cause, there is a sea-level rise and global warming. Sometimes endangered species become disoriented and need a ride home.

As the world population swells and finite resources are competed for, there will be more illegal migrations, fisheries violations and violence among nations. Look at the South China Sea.

Chapter 41

Retirement Finally . . . Sort Of

THE PACE FROM BEING an active-duty captain to a civilian was an adjustment. For the first time since I was a freshman in the Academy, I was now at the very bottom of the organization chart, and I loved it. My new chain of command included my school chief, branch chief, commanding officer, and an admiral (Keith Smith) who had all previously worked for me. Lesson learned here: don't be an asshole, because it could come back to haunt you.

I really don't know how my new civilian colleague Larry Brooks did everything before I arrived. With his marine safety and regulatory background and my experience, we became a great team. From unlocking the building to running a coffee mess to vacuuming classrooms, we did it all. In fact, they combined our names and called us BROGLE because we team-taught the majority of the classes. Our primary students, captains and commanders, filled the top five positions at the thirty-seven sectors. I taught numerous classes filled with students from all over the world. One class of thirty-five were from twenty-eight

Sector Commander Class at a boat simulator

different countries. Additionally, we helped in many other classes as well, even teaching seaman.

Speaking of old captains, retired Captain Robert Holt, US Navy, was the director of the Coast Guard's International Division, which was housed in our building at Yorktown. It had grown from the time I was the team chief of eight to now over fifty deployers. Silver-haired Robert was a year older than Larry and always fun to be around. He had been a destroyer CO during Desert Storm and was VADM Kalleres's operations officer. He was also involved in the investigation of the *Cole* bombing. We had plenty of stories to reminisce on the time I spent with his second fleet and his former boss crossing the Atlantic. His last job was as the executive assistant for the chief of Naval Operations.

After a couple of years into the new job, the European Union (EU) contacted our state department who reached out to Robert and requested a maritime security course be delivered in Brussels. Ever the opportunists, Larry and I raised our hands to support. The course they wanted didn't exist, so we created it.

A funny moment was when we boarded the plane for Brussels. The EU paid for our first-class flights. Neither Larry nor I had ever flown first class. Working for the US government, we were always required to fly in coach. Plus, I'm just cheap. But this time, the EU was paying for it.

Upon boarding the plane, we discovered first-class wasn't just a big reclining seat upfront; it was a pod. We could even lay flat. They even had massage features and a bag of goodies. I was one pod behind Larry. He quickly explored the gift bag, fluffed his pillow, and tried on the booties and mask.

People were still coming in, and then it happened. Just after he got his pre-take-off glass of champagne, an annoyed passenger stopped at his pod. "Sir, you're in my seat."

I gave him crap for being low-class for the next several years.

Larry gave up smoking for his grandkids, but I never fulfilled my new year's resolution to get him to the gym. The last time he was in one was in the mid-1970s at the Academy. He was in pretty good shape and about ten years older than I was. Some of our older colleagues told me not to poke the statue.

Larry got me good once after our European trip. The staff biked down past Hamilton's redoubt #10 to the Yorktown waterfront for a staff breakfast. Some of the older folks drove, including Larry. I bet

Larry, if he gave me a five-minute head start, I could bike back to the base before he could drive there. It was on. I was screaming; my finely tuned legs still buff from my 100-kilometer race without Dean Lee. I made quick work of the first very steep hill. Past redoubt #9 and #10. I kept looking back but no sign of Larry. As I crossed the victory line, there still was no sign of Larry. That asshole called it a day and just went home. Technically, I won.

My daughter Storm was anxious to go to Thailand, so sure enough, I volunteered for a week-long trip back to Bangkok. Unfortunately, her high school schedule wouldn't allow it, but I socked away frequent flyer miles for future, father-daughter adventures.

The Thai people are wonderful, and I'm not just saying that because my wife and mother-in-law are Thai. Before I deployed for the mission to Thailand, I had a Thai captain in a resident course at Yorktown. He described the three types of weather in Thailand: it's either hot, very hot or fucking hot.

Temperature wasn't the only thing hot going on in Thailand.

Just before I arrived, an international drama took place. A Thai kids' soccer team took shelter in a cave during a torrential downpour. The rising water pushed them further and further back in the cave to the point where they were completely cut off and a mile in. The rainy season had started early, meaning they were going to be trapped potentially for months. A huge, cave-rescue effort was mounted to feed in air and food to the kids. Eventually, the Thai Navy SEALs swam in, administered drugs to the kids so they wouldn't panic and got them all out. One Navy SEAL died on this daring rescue mission. In my class, I had five different agencies as students. One of the students was a Thai Navy SEAL who had been part of the rescue.

Kicking off the Sector Commander Course with Admiral Linda Fagan, the first Female Service Chief

The great thing about a training base like Yorktown is you still got to see old shipmates. LT Jason Harris was on the International Staff. We had served together for a year in San Diego at PACTACLET. He had been prior enlisted and was now at the twenty-year mark and retiring. He reached out to his former ship commanding officer, Karl Schultz, to preside over the ceremony. If that name rings a bell, he was currently selected to be the next Commandant. Admiral Schultz agreed to be the retirement speaker, but three days from the event Admiral Schultz had a short notice trip pop up in Africa. Jason asked me to fill in and be the retirement speaker. I never say no, because it's the right thing to do.

This event was going to take place in the auditorium with all the international students. I started diligently crafting up remarks, hoping it would be less stressful than the moments before Sharon Doggett's retirement.

The day before the ceremony, Admiral Schultz called me. "Mark, guess what. My trip to Africa was canceled, so I can still make Jason's retirement."

I thought that was awesome news; I didn't have to give the speech in the auditorium; I can just watch and clap. Nope. Admiral Schultz said he would be in the audience, but I was to give the speech. Terrific!

The speech went fine, although Karl Schultz would have done a much better job, no doubt. His mere presence speaks volumes about our current Commandant and Jason.

Through Jason and Robert Holt, I began to once again support international training on a regular basis. Mexico was recently named the most violent country in the world, even surpassing Syria, which was at war. The drug cartels control roughly two-thirds of the country. The cartels are at war with each other and the government. Corruption is also a huge issue. Because of this, SEMAR (the Mexican Navy) was seen as the least corrupt entity in the government, at least from the maritime perspective. They decided to pursue a similar organizational structure to US Coast Guard sectors.

I know drugs, I know cartels, and I know sectors. Sign me up, coach. To give you a sense of what these students faced, I offered a class on public affairs and media interviews, which is a standard part of the maritime officer course. We had to cancel it for this mission, as a couple of years ago, they had a drug bust with a media component. One of the seizing officer's names was leaked to the media. The cartels assassinated him and his entire family.

During the week we were in the country, there was gunfire around the hotel, and we required armed escorts to and from the base. It felt a lot like my trip to El Salvador.

There is a particularly violent cartel of ex-special forces at work now in Mexico. The other cartels have formed a task force to go after this rouge element as it's bad for the overall business. Thinking back to the strategy of cutting the head off the snake by capturing the cartel leadership may need to be rethought. The cartels have essentially bought off the public. They are more trusted in some cases to provide security than the military and police.

When the head of the cartel is removed, the remaining lieutenants or other cartels often move in to seize power and market share. Crime spikes, and confidence in the government to provide safety is reduced. Perhaps the better move is to target the most violent elements. The war on drugs is a complicated matter, and it's easy for some in the United States to think our partners aren't doing enough. They often risk much more than we do.

I started the training with a slide of a football stadium with 100,000 people.

That's the number of fatalities in just one year in Mexico and the United States due to drugs. Have no doubt; we need to be aligned for the good of both nations.

Drug related deaths in Mexico and U.S. last year. >100,000

After Mexico, other countries requested the Maritime Operations Course. We went to Saudi Arabia twice, Costa Rica, and the United Arab Emirates. Then on to Honduras and Philippines just following Vice President Harris visits; we even had the same interpreters. Then on to the South Pacific with a layover in Australia and training in Vanuatu and Fiji. It was great to travel the world and meet with foreign partners doing the same missions we had in the US Coast Guard. It was also a terrific source of frequent flyer miles!

I flew my daughter down to Costa Rica and week spend a week to remember in that beautiful country.

As a retiree, it's important to volunteer and give back. When we landed in Smithfield in 2009, we knew it would be our home for a while. Both my kids were big into sports, so aligning with the small-town YMCA made sense. I volunteered to serve on the Luter YMCA board. My daughter was active on the swim team, and both my son and daughter were lifeguards for their first jobs. One word of caution when going into any volunteer service, don't miss

Testing out the rope swing in Costa Rican river

meetings! I started as just a member, missed a meeting and was promoted to the youth development division chair. I missed another meeting and was promoted to the board's vice-chair. The third time, I was promoted to the chair.

As I was preparing to retire from active duty, I wasn't looking to pad a resume anymore. The staff there worked wonders though, and those on the board had big hearts. While the staff drove this, one thing I was particularly proud of was our partnership with the local government and providing free swim lessons for every second grader in the county. My daughter Storm became one of the instructors.

That initiative caught on and the regional YMCA in Hampton Roads adopted the program.

Lessons Learned:

When boarding a plane, before settling in, always double check your seat assignment.

The United States is much better off than our European friends when it comes to maritime security.

If you're covering a speech for the Commandant, and he shows up, don't suck

Be careful betting against someone that refuses to go to the gym.

Mexico has great potential and is the fifteenth largest economy in the world, but it will be a fight to retake their country. Success is possible. Just look at Colombia.

As you get older test for balance in your life: go to a funeral. What would your family member, friend and colleague say about you at your funeral. It's never too late to work on this.

Teaching a class of military officers from 21 different countries is rewarding. Sending your brother a message after Monday Night Football - Priceless.

Chapter 42

Fly Like an Eagle

IF YOU'VE NEVER HEARD the lyrics to *Fly Like an Eagle,* do an internet search for it. This hit song by the Steve Miller Band is all about living and time slipping into the future. It speaks to feeding the babies, shoeing the children, and housing the people living on the streets. There's a solution: it's to *Fly like an Eagle.* This was written when I was twelve. I heard it when Chip Shankle and I crossed the country on my first big adventure in Boy Scouts.

Eagle Scout and captain were only benchmarks in my life, not destinations. Goals and principles are particularly important and keep one focused and productive. When contemplating difficult or life-threatening decisions, go back to the principles to ensure you're on the right track for the right reason. I like to do the *60 Minutes* test.

Can you explain your actions during a rigorous TV interview and be proud of what you did and why you did it. It's a quick but effective gut check in the heat of battle. Captain Kristi Luttrell, who had been the sector commander in New Orleans, decided to take commercial and political heat to stop a decades-long leaking oil well. Sometimes doing the right thing initially is painful but is something you can be proud of the rest of your life. She rocked her interview on *60 Minutes.*

As you become more senior in any organization, the decisions and their consequences become more complex. VADM Lee had a real story of a man trapped in an overturned vessel. Two large fishing vessels had gone out together when the weather besieged them. One of the vessels capsized. All but one of the men escaped and were recovered by the companion vessel. They called in the mayday, and a rescue helicopter was dispatched from Atlantic City, New Jersey. When the helicopter's survival swimmer Jake Stall was lowered to the overturned boat, he heard banging and screaming underneath the bobbing upside-down boat.

Just the bow with an air pocket was exposed above the water. There was debris everywhere, from mooring lines to fishing lines with hooks. Alongside Jake were survivors and family members of the trapped man pleading with the swimmer to attempt to save their loved one. The only thing keeping the boat afloat was the air pocket in the bow where the man was trapped and pounding. If Jake attempted to cut a hole, the air would be released, and the vessel would immediately sink.

The Coast Guard's policy is clear: swimmers aren't divers, and you can only put your hand under and see if you can grab the person. That wouldn't work in this case as the trapped man was well inside the cabin at the bow. The swimmer only had a small, emergency, breathing device called a helicopter emergency egress device (HEED), in case the helicopter crashed into the water. That violent action often will roll the helicopter upside down and make egress or escape challenging. The extra few seconds of air in the HEED's bottle could mean the difference in survival. Jake was smart to have brought it with him. Also on Jake's mind was his wife and two young kids. Despite those factors, he came to the Coast Guard for a reason. For events like this.

He asked permission to go under and attempt a rescue. The pilot immediately relayed it up the chain to the sector commander and then eventually RADM Lee, the Fifth District Commander. While that was taking place, the trapped man's pounding slowed and then stopped.

There was no answer from above yet, and Jake had to make the call. He went! He got through some of the riggings and then to the lifeless man trapped in the bow and he grabbed him. His small HEED's air bottle went empty, and during his exit with the man in tow, the Jake got tangled in lines but eventually got free and got the man to the surface. Despite valiant attempts to revive the boater, they were unsuccessful.

RADM Lee faced the strategic decision of punishing the swimmer on charges for violating the policy and risking his own life or giving him a medal for his valor. If he does the latter, what signal does he send to other swimmers and the broader organization about following policy? Just as the on-scene swimmer and the aircraft commander faced a life-threatening dilemma, he too was facing a *60 Minutes* test.

Even for a two-star admiral, the authority for the appropriate medal ultimately rests with the three-star Atlantic Area commander. At that time, the commander was not on board for the medal. So, RADM Lee found a third option. He wrote the swimmer and his family a letter, explaining his predicament, saying that he wanted the family to know that the swimmer had done one of the bravest, selfless things the admiral had ever seen in his entire career. A couple of years later, Dean Lee got his third star and became the Atlantic Area commander, at which time he approved the medal.

In closing, it has been an honor to be a public servant and work with the great men and women of the US Coast Guard for forty years. It's also my observation that our country has recently been through the ringer in the last few years with the covid pandemic, growing threats overseas, exploding crime, deteriorating education, drug epidemics, mental health concerns and homelessness. A very vocal small minority on both the left and right of our political spectrum has all but censored free speech and shutdown open debate on

LCDR Tom Condit and Mark Ogle in Thailand after bravely eating an eyeball

ideas to solve our national problems. Despite all those bad bombarding headlines, normal Americans make up the vast majority of this country and are truly good people who want the best for their kids and grandkids. Among them, we must have brave men and women to step up and risk it all for the right reasons. Those are the future true heroes needed to right the ship. It's basic leadership and almost any problem can be solved. It's truly our generational duty. It's time for the eagles to fly.

Lessons Learned:

An eagle on the ground is vulnerable. The left and right wings must work together to lift the center. Once in the air, the eagle has no rivals.

Our national goal has always been to form a more perfect union. The military's role is not to determine the destination, but to facilitate the journey.

If you ain't failing, you ain't trying.

Life is short. Don't waste it. Have a mission you can be proud of.

Happiness = living, loving, learning, and leaving a legacy. Have a plan to hit all four, taking advantage of all opportunities along the way. This will lead to a fulfilling adventurous life.

Terminology and Ranks

SCOUT TERMINOLOGY

Scout Ranks - Scout, Tenderfoot, First Class, Star, Life, Eagle

Troop - A group or unit of children under a Scout Master made up of patrols

Patrol Leader - A child's position of responsibility for a group of five to eight children within a troop

Senior Patrol Leader - A child's position as the senior scout in the troop

Merit badges - Completion of required tasks earns patches required for promotion within scouts

Order of the Arrow - Special organization within scouts requiring passage of a survival ordeal

Philmont - High adventure camp in New Mexico

COAST GUARD RANKS AND ABBREVIATIONS

Cadet - College student at the Academy (insignia vary based on year and regimental position)

ENS - Ensign or O1 (single bronze bar)

LTJG - Lieutenant Junior Grade or O2 (single silver bar)

LT - Lieutenant or O3 (silver double bars or railroad track)

LCDR - Lieutenant Commander or O4 (bronze oak leaf)

CDR - Commander or O5 (silver oak leaf)

CAPT - Captain or O6 (silver eagle)

RDML - Rear Admiral Lower Half or O7 (one star)

RADM - Rear Admiral Upper half or O8 (two stars)

VADM - Vice Admiral or O9 (three stars)

ADM - Admiral or O10 (four stars)

COMBATANT COMMANDS
(LED BY FOUR-STAR ADMIRAL OR GENERAL)

AFRICOM - Africa Command
CENTCOM - Central Command
CYBERCOM - Cyber Command
EUCOM - European Command
NORTHCOM - Northern Command
PACOM - Pacific Command
SOUTHCOM - Southern Command
SPACECOM - Space Command
STRACOM - Strategic Command
SOCOM - Special Operations Command
TRANSCOM - Transportation Command

Acronyms

ASAT - antisatellite

BM1 - First Class Boatswains Mate

BUD/S - Basic Underwater Demolition/SEAL

CBP - Customs and Border Protection

CO - commanding officer

COCOM - combatant commander

DEA - Drug Enforcement Agency

DHS - Department of Homeland Security

DIAT - Drug Interdiction Assist Team

DOG - Deployable Operations Group

EU - European Union

FEMA - Federal Emergency Management Agency

HAF/BAF - helicopter and boat assault force

HAZMAT - hazardous material

HEED - helicopter emergency egress device

HRT - hostage rescue team

ICE - Immigration and Customs Enforcement

IMLET - International Maritime Law Enforcement Team (formerly DIAT)

IMLETT - International Maritime Law Enforcement Training Team

J3 - chief of operations

LEDETS - law enforcement detachments

MOTR - Maritime Operations Threat Response

MSRT - Maritime Security Response Team

MSST - Maritime Safety and Security Team

NASA - National Aeronautics and Space Administration

NCBT - Noncompliant Boarding Team members

NCW - Naval Coastal Warfare

NFL - National Football League

NRO - National Reconnaissance Office

NSFCC - National Strike Force Coordination Center

OCS - Officer Candidate School

OOD - officer of the deck

OODA - observing, orienting, deciding and acting

OPLAN - operations plan

PACTACLET - Pacific Tactical Law Enforcement Team

PATFORSWA - Patrol Forces Southwest Asia

PSU - Port Security Unit

R&R - rest and relaxation

ROTC - Reserve Officers' Training Corps

SEAL - Navy Sea, Air, and Land

STRATCOM - Strategic Command

TACLET - Tactical Law Enforcement Team

TSA - Transportation Security Administration

UMOPAR - translates to the Mobile Police Unit for Rural Areas, a subsidiary of the Special Anti-Narcotics Force of the Bolivian National Police

UNB - unannounced nighttime boardings

UNC - University of North Carolina

VPIR - Visible Intermodal Prevention and Response

XO - executive officer

Other Book Recommendations

BOOKS THAT DETAIL SOME of these missions from a different perspective:

Mayday Mayday! The Most Exciting Missions of Rescue, Interdiction, and Combat in the 200-Year Annals of the Coast Guard by Samuel Schreiner (St. Croix and Vashon drug busts)

Rescue Warriors: The U.S. Coast Guard, America's Forgotten Heroes by David Helvarg (St. Croix, PACTACLET 40 tons of cocaine, and DOG standup)

Not Your Father's Coast Guard: Untold Stories of Coast Guard Special Forces by Matthew Mitchell (Bolivia and Colombia missions)

The Superferry Chronicles: Hawaii's Uprising Against Militarism, Commercialism, and the Desecration of the Earth by Jerry Mander and Koohan Paik (Kauai Mission)

One Damn Thing After Another: Memoirs of an Attorney General by Bill Barr (St. Croix)

About the Author

BORN IN CHAPEL HILL, North Carolina, Mark Ogle graduated from the US Coast Guard Academy in 1986 and the Naval War College in 2007.

During his career, he traveled to seventy-six different countries and sailed with the Second Fleet, training numerous Navy boarding teams in route to Desert Storm. He also led the White House-ordered armed landing party amidst gunfire to rescue seventy-seven barricaded tourist on the island of St. Croix.

Throughout his command tours and as a boarding officer, he made over sixty narcotics seizures that included the first drug-laden semi-submersible, the first liquid cocaine load and—at the time—the largest maritime bust in history. His San Diego-based unit seized over 130 tons of cocaine worth $3.2 Billion in three years.

Eleven of his twenty-seven active-duty years were in command. He retired as a Coast Guard captain and the chief of Future Operations for the Atlantic Area, where he was responsible for planning, managing resources and providing oversight of all US Coast Guard operational missions which span from the Rocky Mountains east to the India/ Pakistan border.

Following his retirement from active duty in May 2013, he continues to support the Coast Guard as a civilian training specialist and course administrator for the Sector Commander, Sector Department Head, and International Maritime Operations Courses.

Deployments

Operation Snow Cap (South American counterdrug operation)
Operation Hawkeye (operation in St. Croix)
Operation Desert Shield (predecessor of Desert Storm)
Embargo of Serbia (predecessor of Operation Deliberate Force)
Operation Able Vigil (mass migrant exodus from Cuba)
Operation Able Manner (mass migrant exodus from Haiti)
Operation Uphold Democracy (intervention in Haiti following military coup d'état)
Operation Shadow Game (capture of drug cartel leadership)
Operation Burnt Frost (response to deorbiting satellite)
Operation Iraqi Freedom (liberation of Iraq)
Operation Enduring Freedom (war in Afghanistan and larger scale War on Terror)

Personal Awards

1986 Coast Guard Academy Superintendent's Award
2006 United States Interdiction Coordinator Operational Intelligence Award
2012 Virginia Maritime Association Port Champion
2020 Captain Neils P. Thompson Innovation Award
Coast Guard Meritorious Service Medal x 7
Joint Service Commendation Medal
Coast Guard Commendation Medal x 3
Achievement Medal x 2
Commandant's Letter of Commendation
Several unit awards

www.ingramcontent.com/pod-product-compliance
Lightning Source LLC
Chambersburg PA
CBHW060900120626
46553CB00001B/151